S. David Jackson
Hydrogenation

Also of Interest

Biomass and Biowaste.
New Chemical Products from Old
Balu, García (Eds.), 2019
ISBN 978-3-11-053778-9, e-ISBN 978-3-11-053815-1

Electrochemical Energy Systems.
Foundations, Energy Storage and Conversion
Braun, 2018
ISBN 978-3-11-056182-1, e-ISBN 978-3-11-056183-8

Chemical Energy Storage.
Schlögl (Ed.), 2012
ISBN 978-3-11-026407-4, e-ISBN 978-3-11-026632-0

Physical Sciences Reviews.
e-ISSN 2365-659X

Hydrogenation

Catalysts and Processes

Edited by
S. David Jackson

DE GRUYTER

Editor
Prof. Dr. S. David Jackson
University of Glasgow
School of Chemistry
Joseph Black Building
University Avenue
Glasgow G12 8QQ
United Kingdom
David.Jackson@glasgow.ac.uk

ISBN 978-3-11-054373-5
e-ISBN (PDF) 978-3-11-054521-0
e-ISBN (EPUB) 978-3-11-054375-9

Library of Congress Cataloging-in-Publication Data
Names: Jackson, S. D. (S. David), editor.
Title: Hydrogenation : catalysts and processes / edited by S. David Jackson.
Description: Berlin ; Boston : De Gruyter, [2018] | Includes bibliographical references and index.
Identifiers: LCCN 2018030272 (print) | LCCN 2018036650 (ebook) | ISBN 9783110545210 (electronic Portable Document Format (pdf)) | ISBN 9783110543735 (print : alk. paper) | ISBN 9783110543759 (e-book epub) | ISBN 9783110545210 (e-book pdf)
Subjects: LCSH: Hydrogenation.
Classification: LCC QD281.H8 (ebook) | LCC QD281.H8 H9275 2018 (print) | DDC 547/.23–dc23
LC record available at https://lccn.loc.gov/2018030272

Bibliographic information published by the Deutsche Nationalbibliothek
The Deutsche Nationalbibliothek lists this publication in the Deutsche Nationalbibliografie; detailed bibliographic data are available on the Internet at http://dnb.dnb.de.

Typesetting: Integra Software Services Pvt. Ltd.
Printing and binding: CPI books GmbH, Leck
Cover image: dkidpix/iStock/Getty Images Plus

www.degruyter.com

Preface

In 1912 Paul Sabatier was awarded the Nobel Prize in Chemistry for *"for his method of hydrogenating organic compounds in the presence of finely disintegrated metals whereby the progress of organic chemistry has been greatly advanced in recent years"*, almost 100 years later in 2007 Gerhard Ertl was also awarded the Nobel Prize in Chemistry for *"for his studies of chemical processes on solid surfaces"* related in no small part to his work on nitrogen hydrogenation to ammonia. Today the hydrogenation catalyst market is worth ~$3 billion and hydrogenation is used in all areas of today's chemical industry, from refinery to pharmaceutical processes. However the ability to run large hydrogenation processes does not mean that we understand the chemistry on the molecular level. Indeed with many processes only a rudimentary understanding of the fundamental chemistry has been achieved. There is much to learn and understand in catalytic hydrogenation.

The literature on hydrogenation is vast and this book makes no attempt to cover all aspects. Instead, in the following chapters, you will find a selection of topics covering hydrogenation catalysts and processes written by leading experts in their fields. In chapter 1 Martin Lok gives an overview of hydrogenation, looking at the role of the active metal, promoters and supports. In chapter 2 Steve Schmidt explores the Raney catalyst system from genesis, through various developments and applications to future opportunities. In chapter 3 Swetlana Schauermann examines in detail the interaction between reactant and catalyst and reveals the complexity in the hydrogenation of olefins and ketones. Chapter 4 discusses the hydrogenation of aromatics, a perfect example of well-developed industrial processes that do not have the depth of fundamental understanding. In chapter 5 Alan Allgeier and Sourav Sengupta give an in depth exposition of nitrile hydrogenation, examining both fundamental and applied aspects for heterogeneous and homogeneous catalysts. In chapter 6 Gordon Kelly addresses Fischer-Tropsch synthesis focussing on cobalt systems, discussing reactors and catalysts. In the final chapter Justin Hargreaves takes a fresh look at ammonia synthesis examining nitrides and other new catalyst opportunities.

Hydrogenation has been an active area of research and application for more than 100 years, this book makes it clear that it still remains so.

Finally let me thank the contributors to this book and the publisher for all their efforts – you made my job easy.

S David Jackson
Glasgow 2018

https://doi.org/10.1515/9783110545210-201

Contents

List of Contributors

Ahmed K. A. AlAsseel
University of Glasgow
School of Chemistry
Joseph Black Building
University Avenue
Glasgow G12 8QQ
United Kingdom
a.alasseel.1@research.gla.ac.uk

Alan M. Allgeier
University of Kansas
Chemical and Petroleum Engineering
and
The Center for Environmentally Beneficial Catalysis
1530 W. 15th St
Lawrence, KS 66045
alan.allgeier@ku.edu

Justin S. J. Hargreaves
University of Glasgow
School of Chemistry
Joseph Black Building
University Avenue
Glasgow G12 8QQ
United Kingdom
Justin.Hargreaves@glasgow.ac.uk

Gordon J. Kelly
Johnson Matthey
PO Box 1, Belasis Avenue
Billingham, TS23 1LB
United Kingdom
Gordon.Kelly@matthey.com

Kathleen Kirkwood
University of Glasgow
School of Chemistry
Joseph Black Building
University Avenue
Glasgow G12 8QQ
United Kingdom
k.kirkwood.1@research.gla.ac.uk

C. Martin Lok
Catalok Consultancy
Vliegenvangerlaan 12
2566 RN Den Haag
The Netherlands
m2lok@casema.nl

Swetlana Schauermann
Christian-Albrechts-University Kiel
Institute of Physical Chemistry
Max-Eth-Str. 2
24118 Kiel
Germany
schauermann@pctc.uni-kiel.de

Stephen R. Schmidt
Grace Davison
7500 Grace Drive
Columbia, MD 21044
USA
Steve.R.Schmidt@grace.com

Sourav K. Sengupta
DuPont Safety & Construction
Experimental Station
200 Powder Mill Road
Wilmington, DE 19803
Sourav.Sengupta@dupont.com

https://doi.org/10.1515/9783110545210-202

C. Martin Lok

1 Structure and performance of selective hydrogenation catalysts

1.1 Introduction

Hydrogenation is one of the major reactions for the synthesis of pharmaceuticals, agrochemicals, fine chemicals, flavors, fragrances, and bulk chemicals like methanol and ammonia. The reaction generally is highly selective and easy to work up. It is commonly employed to reduce or saturate organic compounds in the presence of a catalyst such as Ni, Pd, or Pt. Rylander called it "one of the most powerful weapons in the arsenal of the synthetic organic chemist" [1]. An important feature is the atom economy, which for several reactions is 100% if the reaction is carried out by hydrogen, the cleanest of all reducing agents. The process usually does not generate any waste provided that the catalyst is selective and the excess hydrogen (and solvent) can be recycled. Moreover, heterogeneous catalysts can be recovered for reuse or for metal reclamation.

The earliest hydrogenation is that of the Pt-catalyzed reaction of hydrogen with oxygen that was commercialized in the Döbereiner's Lamp as early as 1823. Later in 1897, the French scientist Sabatier discovered that in a gas-phase reaction, Ni catalyzed the addition of hydrogen to hydrocarbons [2]. Shortly afterward, in 1902, Normann was awarded a patent for hydrogenation in the liquid phase [3]. This was the beginning of edible oil and fatty acid hydrogenation which now is a worldwide industry. Normann first used Ni catalysts in the "hardening" of liquid oleic acid to the more valuable solid stearic acid and subsequently applied these catalysts in the hydrogenation of oils and fats. He explored commercialization of the process, first at Joseph Crosfield & Sons in the United Kingdom, and later in Ölwerke Germania in Germany [4]. The catalysts consisted of finely dispersed Ni, supported on, at first, pumice and later on a kieselguhr carrier. In 1926, Murray Raney, when involved in the hydrogenation of cottonseed oil, made his classical discovery of a catalyst based on a Ni/Al alloy [5]. Despite having been used for over 80 years, Raney® or "sponge" catalysts are still essential for a wide variety of industrial applications, such as the manufacture of sorbitol, sulfolane, fatty and alkylamines, hexamethylenediamine, 1,4-butanediol, and various fine chemicals and pharmaceuticals.

The early twentieth century was a landmark period for industrial catalysis and within decades, major processes for the production of methanol, ammonia, and liquid hydrocarbons, all based on hydrogenation processes, were discovered. The commercially important Haber–Bosch process for production of ammonia, first described in 1905, involves hydrogenation of nitrogen allowing the large-scale production of fertilizer. In 1913, Mittasch and Schneider patented the conversion of

https://doi.org/10.1515/9783110545210-001

mixtures of carbon monoxide and hydrogen in the presence of heterogeneous catalysts such as supported Co resulting into the formation of liquid hydrocarbons in a reaction now commonly described as the Fischer–Tropsch (FT) process [6]. This discovery was not immediately followed up because priority was given to the commercialization of the methanol and ammonia processes [7, 8]. Since then, hydrogenation is widely applied for a variety of compounds.

Gaseous hydrogen is by far the most common source of hydrogen and is produced industrially from hydrocarbons by steam reforming or, on a smaller scale, by electrolysis. In organic synthesis, transfer hydrogenation is also used for hydrogenation of polar unsaturated substrates, such as ketones, aldehydes, and imines from donor molecules such as hydrazine, formic acid, isopropanol, and dihydronaphthalene [9]. These hydrogen donors undergo dehydrogenation to, respectively, nitrogen, carbon dioxide, acetone, and naphthalene.

Hydrogenation is a strongly exothermic reaction. The hydrogenation of an alkene involves a Gibbs free energy change of -101 kJ.mol^{-1}. Even for the partial hydrogenation of a large molecule as a triglyceride (MW 890), the heat generated can be sufficient to raise the temperature of the oil by 50–100 °C. Bulk chemicals hydrogenation can be done in the gas or in the liquid phase. Mostly, no solvents are used.

In 1934, Horiuti and Polanyi proposed a reaction scheme assuming that the hydrogenation of alkenes occurs in three steps [10]:

(a) Binding of both the unsaturated bond and the hydrogen molecule onto the catalyst surface. The hydrogen dissociates into atomic hydrogen.
(b) Addition of one atom of hydrogen to the adsorbed olefin. In this step, the intermediate formed can rotate and, after releasing a hydrogen atom, may detach from the catalyst surface.
(c) Addition of the second atom; this step is irreversible under hydrogenating conditions.

Free rotation of alkyl groups in the half-hydrogenated intermediate state results in *cis–trans* isomerization. At low hydrogen pressure, the rate of isomerization can even be higher than the hydrogenation rate. In addition, a double bond shift is possible by H abstraction from an adjacent CH_2 group. These are general phenomena in partial hydrogenation.

The main classes of hydrogenation catalysts are homogeneous catalysts, stabilized metal nanoparticles, and heterogeneous catalysts. Homogeneous catalysts are often based on platinum group metals (PGMs), e.g., Wilkinson's catalyst, RhCl $(PPh_3)_3$. Homogeneous catalysts are superior in asymmetric synthesis by the hydrogenation of prochiral substrates. An early demonstration of this approach was the Rh-catalyzed hydrogenation of enamides to produce the drug L-DOPA. In principle, asymmetric hydrogenation can be catalyzed by chiral heterogeneous catalysts too, but selectivity often is inferior to that of homogeneous catalysts. Homogeneous and enantioselective hydrogenations have recently been reviewed by De Vries et al. [11]

and Blaser et al. [12–14]. The current chapter focuses exclusively on heterogeneous catalysts which are the catalysts most commonly used in commercial hydrogenation. For recent reviews on general hydrogenation, see Refs. [15–19].

1.2 Selective hydrogenations

Typical substrates for hydrogenation are alkenes, alkynes, aldehydes, ketones, esters, carbon monoxide, nitriles, and nitro-compounds which are converted into the corresponding saturated or partial-hydrogenated compounds, i.e., alkanes, alkenes, alcohols, and amines. Often complete saturation has to be avoided and the reaction should be stopped in, e.g., the mono-ene stage for polyolefins. Almost always a single hydrogenation product is required.

Examples of selective hydrogenation are as follows:

- Partial hydrogenation of vegetable oils consisting of esters of glycerol and long-chain mono-enoic, di-enoic, and tri-enoic fatty acids. Depending on the target product, the reaction has to be terminated in the mono-ene or di-ene stage.
- Production of pyrolysis gasoline. Selective hydrogenation of di-enes to mono-enes without saturating the aromatics.
- Hydrogenation of triple bonds while avoiding full saturation. Examples are the removal of acetylene in ethylene and the partial hydrogenation of phenylacetylene to styrene.
- Conversion of alkyl nitriles into primary amines without formation of secondary and tertiary amines.
- Unsaturated carbonyl hydrogenation to unsaturated alcohols while minimizing saturated aldehydes and saturated alcohols.
- Carbon monoxide and carbon dioxide hydrogenation to exclusively methane, methanol, or alkanes.
- Hydrogenation of benzene to cyclohexane while avoiding formation of by-products like methylcyclopentane, a.o.
- Asymmetric hydrogenation to chiral compounds.

In general, the requirements for a heterogeneous hydrogenation catalyst are high activity, excellent selectivity, long lifetime and reusability, and/or recyclability of the catalyst. In addition, fast filtration for powder catalysts is required. Activity and selectivity are strongly dependent on the choice of the main active metal. The main metal influences the strength of adsorption of reactants, the rate of desorption of reaction products, and the rate of chemical transformations. Metals most frequently used in heterogeneous catalytic hydrogenation are Pd, Pt, Rh, Ru, Ni, Co, and Cu. More rarely, Ir, Os, and Re are used. Subsequently, promoters (or selective poisons), additives, and supports have to be selected. For some reactions, selection of the carrier with the right porous structure and particle size distribution can be extremely critical. In addition, selectivity often is strongly dependent on the selection of reactor type and process conditions like temperature and pressure. Thus in the rhodium-catalyzed hydrogenation of chloronitrobenzenes, higher pressures favor the selective reduction of the nitro group, while at low pressure, the hydrogenolysis of the carbon–chlorine bond prevails [20, 23]. Hydrogenations are often run neat. In some cases, solvents are used to dissolve the substrate or to moderate reactions or exotherms. The effect of solvent is often reaction specific. Common solvents are water, methanol, ethanol, ethyl acetate, cyclohexane, hexane, acetone, a.o.

1.3 Main active metal

The main metals in industrial hydrogenation are the precious metals Pt, Pd, Rh, and Ru (PGMs) and the base metal catalysts Co, Cu, and Ni. PGMs form highly active catalysts which may operate under mild conditions while the base metals generally are much less active and therefore require higher metal loadings and/or more severe conditions. While supported PGM catalysts commonly only contain 0.5–5 wt% metal, base metal levels range from 30 to sometimes close to 100 wt%. The trade-off is activity versus cost. As shown in Table 1.1, price is a major consideration.

Table 1.1: Metal prices in $\$.kg^{-1}$. Adapted from Refs. [21, 22].

Metal	Price ($\$.kg^{-1}$)
Pt	31,000
Rh	31,000
Pd	25,000
Ru	1,300
Co	53
Ni	10
Cu	6

Most PGMs and base metals in heterogeneous catalysts are recycled. Catalysts can be returned to the manufacturer or metal reclaimer after use for refining and remanufacturing, allowing the intrinsic value of the metal to be recovered. The efficiency of metal recovery during refining varies from metal to metal (Table 1.2) and some losses are inevitable. Especially for Ru, the recovery rate is rather low because the RuO_4 formed during calcination of the spent catalyst is quite volatile.

Probably the most important consideration in selecting a metal is selectivity. A high selectivity will ensure a high yield and will minimize purification costs. Though for each application several metals may be suitable, in practice often only one or two specific metals are favored. Thus, Pd shows extraordinary chemoselectivity in the

Table 1.2: Typical losses in refining of platinum group metals used as hydrogenation catalysts [23].

Metal	Losses (%)
Pd	2–5
Pt	2–5
Rh	5–10
Ru	10–15

partial hydrogenation of alkynes, superior to all other metals. A drawback of Pd is that it exhibits strong double bond migration and isomerization, which may lead to undesired side products. In general, Pd is the most versatile of the PGMs and is preferred for the hydrogenation of alkynes, carbonyls in aromatic compounds, etc. Pt is the preferred metal for the hydrogenation of halonitroaromatics and reductive alkylations. Rh catalyzes ring hydrogenation of aromatics while Ru is used for aliphatic carbonyl hydrogenation.

Hydrogenation of acetylene in ethylene is an important industrial process in which the balance between semi-reduction and over-reduction is extremely critical. Small amounts of acetylene in ethylene from naphtha crackers have to be removed by selective hydrogenation to usually <1 ppm acetylene because acetylene deactivates the catalyst for ethylene polymerization and inhibits chain growth and thus reduces molecular weight of the polymer [19, 20]. Of the PGMs, Pd shows by far the highest selectivity and Pd/alumina is the preferred catalyst, at least for cleaner feeds. The presence of CO improves selectivity in front-end hydrogenation. Because of its greater poison resistance, sulfur-promoted Ni/alumina catalysts may be used for dirtier ethylene feeds.

For the removal of both methylacetylene and propadiene from the C3-cut, or vinylacetylene and butadiene from the C4-cut, of a steamcracker feed, Pd/alumina is used too. Butadiene is hydrogenated on Pd catalysts with a selectivity close to 100%, whereas on Pt-catalyzed hydrogenation under comparable conditions is much less selective and leads to a considerable formation of butane [24, 25]. This effect may be related to the stronger di-σ-adsorption on Pt compared to Pd which increases the chance of subsequent hydrogenation.

In the processing of triglycerides in edible oils, polyunsaturates are hydrogenated to mono- or di-olefins in a batch slurry process. In this process, Ni is used exclusively. Ni has the benefits of combining high selectivity and low cost. This used to be and probably still is the largest single application of hydrogenation [26]. A complicating reaction is *cis–trans* isomerization leading to *trans* fatty acids [27]. PGMs are not used at all because of undesirable side reactions and costly metal losses during filtration and subsequent work up. Similarly, in the related hydrogenation of unsaturated fatty acids, mostly powdered Ni catalysts are used but Pd catalysts sometimes are used too. These catalysts are applied in a continuous (slurry bubble column) or, more commonly, in a batch slurry process. Because of the corrosive nature of fatty acids, Ni catalysts in fatty acid hydrogenation rapidly deactivate by loss of Ni dispersion and because a considerable amount of Ni readily dissolves as a Ni acylate which eventually ends up in the distillation residues (see also Section 1.5). Therefore, some fatty acid manufacturers prefer to use the not only much more expensive but also much more acid-resistant Pd, especially for high-molecular weight dimeric acids that are difficult to distil.

Copper generally exhibits a very low catalytic activity but because of its exceptional selectivity, it is the metal of choice in the high-volume production of methanol from syngas. In addition, Cu is the main metal in carboxylic acids/esters hydrogenation at 200–400 bar and 250–300°C to saturated alcohols [28]. The catalyst of choice is Ba-doped copper chromite. Recently, several studies report more environmental friendly non-Cr–Cu catalysts like bi-metallic Cu–Fe catalysts [28]. A main challenge is to develop catalysts that can operate under milder conditions, see also Section 1.4.

Copper is also used commercially to selectively hydrogenate aldehydes to the corresponding alcohols in a gas-phase reaction. Nickel is more active in this reaction but forms by-products like alkanes and, therefore, is mainly used in liquid-phase polishing of alcohols formed earlier by Cu-catalyzed hydrogenation of aldehydes.

Copper was widely studied in the hydrogenation of edible oils. It proved to be more selective than Ni in reducing the tri-enoic content while maintaining the di-enoic fatty acids [27]. In the end, Cu was never commercialized for this reaction mainly because remaining traces of Cu in the filtered oil promoted autoxidation leading to deterioration of the oil.

Zinc is mainly used as co-metal/support in Cu-catalyzed methanol formation but it shows unique properties in its own right as zinc chromite by preserving unsaturation in converting unsaturated carboxylic acids/esters to the corresponding unsaturated alcohol [29].

The most important application of Co catalysts is the high-volume production of alkanes from syngas in the FT reaction [30]. Cobalt-catalyzed FT is characterized by a high C5+ selectivity and a low tendency to form olefins and carbon dioxide. This reaction is being applied commercially using both slurry-bubble column and fixed-bed technology. An alternative to cobalt is iron which is less active and produces more carbon dioxide than cobalt. Similar to cobalt catalysts, precipitated iron catalysts produce alkanes in a low-temperature process while high levels of olefins and oxygenates are produced at higher temperatures using fused Fe catalysts [30].

Compared to Ni and other metals, Co shows a remarkable selectivity toward primary amines in the hydrogenation of alkyl nitriles. Nitriles are reduced stepwise to the imine and eventually the amine. Primary amines may further react to their corresponding secondary and tertiary amines while releasing ammonia. Addition of ammonia helps to suppress secondary and tertiary amine formation and its presence is essential for Ni catalysts. Co catalysts allow the processing at reduced ammonia pressures or even in absence of ammonia. Thus, isophorone diamine, a precursor to the polyurethane monomer isophorone diisocyanate, is produced from isophorone nitrile wherein Co-catalyzed hydrogenation converts both the nitrile into an amine and the imine formed from the aldehyde and ammonia into another amine [31, 32]. The chemoselective hydrogenation of cinnamonitrile to 3-phenylallylamine proceeds with up to 80% selectivity at conversions of >90% with Raney Co but only up to 60% selectivity with Raney Ni [33].

Rh has a high activity for the hydrogenation of aromatic compounds. Its activity for alkene hydrogenation is rather low compared to Pt or Pd. Ru is applied to the hydrogenation of aromatic rings and carbonyl functions. It has the lowest hydrogenolytic activity of the transition metals and is applicable when high selectivity is required. It is used commercially in the production of sorbitol from glucose. Ru is a good FT catalyst but is not used commercially as the main metal, probably for cost reasons. Ir, Re, and Os are only of minor importance as hydrogenation catalysts.

Sometimes there can be a synergistic effect from combining two metals, e.g., monometallic Au catalysts show a high selectivity in the selective hydrogenation of butadiene in excess alkene but a rather low activity. The addition of a controlled amount of Pd to Au nanoparticles induces enhanced catalytic activity compared to monometallic Au catalysts without loss of selectivity. The key parameter to achieve higher activity while maintaining the high selectivity of Au catalysts is the presence of isolated Pd atoms only, and not that of Pd ensembles, onto the Au particle surface [34–35].

The active phase of hydrogenation catalysts usually is the zero-valent metallic state. The reaction typically is assumed to be structure insensitive, i.e., the reaction is independent of size and shape of the metal crystallites. However, several examples have been reported in which hydrogenation reactions are dependent on crystallite size or exposed crystal plane. Thus, a significant production of propenol is observed in the hydrogenation of acrolein over 12 nm Pd nanoparticles, while smaller 4 and 7 nm nanoparticles did not produce any propenol at the temperatures investigated [36]. In the hydrogenation of crotonaldehyde over Pt/SiO_2, the selectivity to the unsaturated crotylalcohol increases with increasing Pt particle size [37]. For large metal particles, the high fraction of Pt(111) surfaces is concluded to favor the adsorption of crotonaldehyde via the carbonyl bond. On small Pt particles, the high abundance of metal atoms in low coordination allows unconstrained adsorption of both double bonds [37]. Apparent structure sensitivity during liquid-phase hydrogenation of citral was attributed to an inhibiting side reaction occurring concurrently with the hydrogenation reaction [38]. The direct hydrogenation of phenol, however, was largely insensitive to Ni particle size for catalysts with a wide range of Ni loadings (0.7–20.9 wt% Ni) and an average Ni particle size from 1.4 to 16.8 nm, albeit there was a discernible decrease in phenol conversion over average Ni diameters less than ca. 3 nm. Hydrogenation selectivity remained unchanged and cyclohexanol was the predominant product irrespective of Ni loading or synthesis route [39]. In the dechlorination of chlorophenols, specific chlorine removal was consistently greater with an increase in the average Ni particle diameter [39]. Overall, it may well be that seemingly facile reactions become structure sensitive when particle size is decreased sufficiently below a few nanometers.

1.4 Promoters

Most heterogeneous catalysts contain promoters or modifiers, which help to fine tune the catalyst for specific reactions. A promoter is defined here as an additive, which itself is mostly inactive but which enhances the activity/selectivity and/or stability and lifetime and/or processability of the metallic catalyst. The promoter can change the geometric and/or the electronic structure of the metal surface and may be chemisorbed onto sites active for unwanted reactions. Commercial catalysts tend to contain multiple promoters. Major exceptions are Ni/Al_2O_3 catalysts prepared by impregnation, which usually are promoter free.

Magnesium increases the dispersion of Ni/SiO_2 catalysts, thus enhancing activity without significantly influencing selectivity. Indeed, co-precipitated nickel catalysts often contain Mg as an efficient way to boost activity [40]. Adding organics like mannitol to a cobalt nitrate solution prior to impregnation is an alternative method to increase metal dispersion of Co/alumina catalysts. These organic compounds react with cobalt nitrate, forming a foam. The structure of the foam is retained in the final calcined product and this effect is responsible for the increased dispersion [41].

Reduction promoters are widely used for Co and to a lesser extent for Ni and Fe. In hydrogenation, Co and Ni operate in the metallic state and relatively high temperatures of 350–500°C are required to activate the oxidic catalyst. Even pre-reduced and passivated catalysts consisting of essentially metallic Co or Ni crystallites, protected with a mono- or bilayer of metal oxide, still require temperatures of 100–200°C to be reactivated. Copper is applied as a reduction promoter for Ni-catalyzed hydrogenation and in Fe-catalyzed syngas conversion. The Sasol low-temperature FT catalyst, originally developed by Ruhrchemie, is promoted by both Cu and K [30]. A typical catalyst contains 25 g SiO_2, 5 g Cu, and 5 g K_2O per 100 g Fe. The Sasol Fe catalyst for high-temperature operation (high-temperature FT) also contains K as a promoter in addition to Mg or Al, but a reduction promoter is not required [30]. Raney-type catalysts do not need to be activated and are ready to use, even at low temperatures [5].

The presence of alkali in iron FT catalysts causes an increased 1-alkene selectivity, a slightly increased reaction rate, an increased growth probability of hydrocarbon chains, and also an increased resistance against oxidation of Fe by the reaction product water [42]. Similarly, alkali addition to cobalt catalysts leads to an increased 1-alkene selectivity. However, in this case, the reaction rate is markedly reduced.

Alkali promotion in Ni-catalyzed nitrile hydrogenation also favors selectivity toward primary amines [43]. The basic character of Raney Ni in this reaction may explain its higher selectivity compared to supported Ni catalysts. Sodium promotion also increases selectivity in Ni-catalyzed hydrogenation of butanal to butanol [44].

Adding reduction promoters like Pt, Re, Ru, and Ir to cobalt catalysts may lower activation temperature by 100–150°C [30, 45]. Apart from facilitating activation, there are additional benefits of increasing active metal surface area by 10% or more [45] and increased stability of the catalyst [46, 47]. Apart from facilitating activation of Co, reduction promoters also increase activity as well as C5+ selectivity. A further advantage is that promotion helps the regeneration of Co catalysts by hydrogen at FT synthesis temperatures while un-promoted Co catalysts cannot be regenerated in this manner [30]. Addition of non-reducible metal oxides such as B, La, Zr, and K causes the reduction temperature of Co species to shift to higher temperatures, resulting in a decrease in the degree of reduction. Finally, some elements promote reduction but either poison the surface of cobalt (Cu) or produce excessive light gas selectivity (Cu and Pd). The presence of copper also increases the selectivity to higher alcohols by an order of magnitude [48]. Both characterization and catalytic studies suggest formation of bimetallic Cu–Co species and enrichment of the surface of the bimetallic particles with Cu [48]. Computational studies indicate that certain promoters may inhibit polymeric C formation by hindering C–C coupling [47].

Particularly for alkyne semi-hydrogenation, catalyst selectivity is even more important than activity. Thus, Pd catalysts are typically used with selectivity enhancing promoters as in the well-known Lindlar catalysts Pd–Pb/$CaCO_3$. Transition metals such as Ag, Rh, Au, Cu, Zn, Cr, and V along with alkali and alkaline-earth elements have been reported. Additives like Ag are used to suppress undesired side reactions such as green oil formation. More recently, lanthanides and especially Ce have been claimed as effective promoters [49].

The addition of carbon monoxide further reduces the consecutive hydrogenation of alkenes [50]. At about 50 ppm CO, a marked increase in ethylene is observed without a significant reduction in ethyne conversion. This has been interpreted as the blocking of sites for ethylene hydrogenation or as competition between CO and H_2 [51, 52]. For modern catalysts, carbon monoxide is not required and, in fact, must be removed to avoid deactivation [53].

Addition of ammonia in edible oil and fatty acid hydrogenation selectively converts polyunsaturated fatty acid groups without any formation of saturated fatty acid groups at all, while relatively few *trans*-isomers are formed. Moreover, linolenic acid, a tri-enoic acid, is hydrogenated more readily than linoleic acid, a di-enoic acid [54].

Basic additives also accelerate the Pd/carbon-catalyzed hydrogenolysis of carbon–halogen bonds and enable the partial hydrogenation of phenols to cyclohexanones. The hydrogenolysis of benzylic alcohols by Pd/carbon catalysts is accelerated by acids and suppressed by bases [55].

As discussed above, Pd/alumina often is the preferred catalyst for removal of acetylene from ethylene but because of their greater poison resistance, sulfur-promoted Ni/alumina catalysts are used for dirtier feeds. These catalysts show an order of magnitude higher tolerance for heavy metals and sulfur than palladium catalysts. Sulfur-promoted Ni catalysts are commercially available or can be produced *in situ* during start-up [56, 57]. Some sulfur in the feed is required to maintain the optimum sulfur content.

Similarly in pyrolysis gasoline hydrogenation, both Pd/alumina and sulfur-promoted Ni catalysts can be used. Pyrolysis gasoline contains a high content of unsaturated hydrocarbons (olefins and aromatics) and must be hydrogenated to eliminate acetylenes, dienes, and aromatic olefins in order to improve the color, reduce gum content, and to reduce fouling in the downstream hydrodesulfurization unit. In general, it is advised to use pre-sulfided Ni catalysts as insufficiently sulfided Ni may give rise to an exotherm resulting in a temperature runaway with temperatures over 750°C. This runaway has led to serious explosions [58]. Presulfiding the catalyst prior to start up will minimize this risk of a runaway and improves the intrinsic safety of the process. A variety of sulfiding agents like H_2S, tetrahydrothiophene, dimethyldisulphide, polysulphides, etc. may be used. Several sulfiding methods exist: (a) reduction of the oxidic catalyst followed by sulfiding, (b) simultaneous reduction and sulfiding of the oxidic catalyst, and (c) sulfiding of the oxidic catalyst followed by reduction. Temperature Programmed Reduction (TPR) studies showed that presulfided oxidic catalysts can be reduced at a lower temperature than the original non-sulfided catalyst [56, 57]. A similar selectivity as for sulfur-promoted nickel catalysts was realized by adding Sn to Ni in a proportion such that the Sn/Ni molar ratio is in the range 0.01–0.2 [59].

The property of a sulfur-promoted Ni catalyst to maximize *cis–trans* isomerization of double bonds while preventing formation of saturates is utilized commercially in the hydrogenation of edible oils to produce cocoa butter replacers [27]. Unlike non-promoted nickel catalysts, sulfur-promoted Ni catalysts produce fats with steep melting curves, thus mimicking the melting behavior of natural cocoa butter.

An interesting concept in sulfur promotion of Ni catalysts is ensemble control. Addition of small amounts of sulfur reduces ensemble size [60, 61]. This technique is applied in the selective hydrogenation of benzene to cyclohexane. Sulfiding of Ni suppresses the formation of by-products like linear alkanes and methylcyclopentane. The rationale is that for hydrogenation, a small ensemble of a few Ni atoms suffices, while for hydrogenolysis to linear alkanes, 15–22 Ni surface atoms are required [61]. The saturation concentration for sulfur is 0.4 and 0.5 sulfur per surface Ni atom for, respectively, the (111) and (100) surfaces. In general, a fully sulfurized Ni catalyst is inactive but this becomes gradually active under sulfur-free hydrogen by stripping of some of the sulfur as H_2S. To maintain selectivity, the feed should contain a low amount of sulfur.

The production of unsaturated alcohols that are important intermediates in the pharmaceutical and flagrance industries involves the selective hydrogenation of unsaturated carbonyl intermediates as a critical step. The hydrogenation of α,β-unsaturated carbonyls into saturated carbonyls is comparatively easy to achieve because thermodynamics favor the hydrogenation of the C=C bonds by 35 $kJ.mol^{-1}$ [62]. Thus, acrolein hydrogenation over Pt yields over >99% C=C and only <1% C=O hydrogenation. In some hydrogenation reactions, like in the hydrogenation of citral, the monometallic supported catalysts are very selective, whereas selective hydrogenation of acrolein and crotonaldehyde cannot be achieved over conventional supported monometallic catalysts [63].

Over Pt, the selectivity to unsaturated alcohols is greatly improved by alloying with Sn, Zn, or Fe because the hydrogenation of the C=O group is promoted and adsorption modes favorable to C=C hydrogenation are inhibited. Several explanations for the high chemoselectivities in carbonyl bond hydrogenation over Pt–Sn are being debated, e.g., alloy formation, presence of ionic Sn, surface enrichment of Sn on Pt, change in the geometry of the Pt particles as well lowered hydrogen adsorption capacity [63]. Recently, inexpensive Ni–Sn-based alloy catalysts, both bulk and supported, were shown to exhibit high selectivity in the hydrogenation of a wide range of unsaturated carbonyl compounds and to produce almost exclusively unsaturated alcohols [64, see also 59]. In phenol hydrogenation, alkali-promoted palladium catalysts show a higher reaction rate of phenol conversion and yield of cyclohexanone than the unpromoted catalyst [65].

The hydrogenation of glycerol to 1,3-propanediol is a challenging reaction as the formation of 1,2-propanediol, which is much less valuable than the 1,3-isomer, is thermodynamically favored. Many studies report a high selectivity toward the 1,2-isomer combined with an almost negligible formation of the 1,3-isomer. In a study by Nakagawa et al., however, glycerol was directly hydrogenolyzed into 1,3-propanediol over a ReO_x-modified iridium catalyst supported on SiO_2 [66–68]. The maximum yield obtained was 38% and the selectivity toward 1,3-propanediol was 68%. Pt–W and Ir–Re catalysts appear to be so far the only effective catalysts with a high selectivity to 1,3-propanediol. The key is to quickly hydrogenate 3-hydroxypropanal before its interconversion to the more stable acetol [69].

Significant progress has been made in carboxylic acid hydrogenation to alcohols under mild conditions (ca. 100°C/20–30 bar) using bifunctional catalysis, The severe conditions (250–300°C/200–400 bar) of current commercial systems, based on copper catalysts and non-functionalized carboxylic acids, lower selectivity in the reduction of highly functionalized or less stable substrates. Therefore, the development of catalysts that display high chemoselectivity at reduced temperatures and pressures is highly desired. New catalysts have been developed, which involve two or more metals in which one metal facilitates heterolytic H_2 cleavage and hydrogenation steps, while the second metal activates the carbonyl group of the acid/ester molecule [70]. Catalysts combining hydrogenation (e.g., Ru, Pt, and Pd) and promoter (e.g., Sn, Re, and Mo) metals show substantial synergy. In addition, support materials such as TiO_2 and ZnO may assist in carbonyl group and/or hydrogen activation [70]. In the hydrogenation of unsaturated carboxylic acids over the bimetallic Ru/Sn catalysts, there is the additional benefit of near total suppression of the hydrogenation of C=C bond in favor of the activation of the hydrogenation of the carboxylic bond, thus leading to the selective formation of unsaturated alcohols [71]. First examples of such catalysts utilized Sn as a promoter for Ru-catalyzed hydrogenation of carboxylic acids and their esters. In a proposed reaction mechanism, metallic Ru sites promote the H_2 dissociation while the adjacent Sn^{2+}/Sn^{4+} Lewis acid sites polarize the carbonyl group of the ester [70, 71].

Few ternary and quarternary metal systems have been explored. Nevertheless, some multi-promoter catalysts have been published, mostly in the patent literature, e.g., for the hydrogenation of hexamethylenedinitrile to the corresponding diamine, a fused Fe catalyst with a range of promoters was claimed (wt%): Fe 72, Mn 0.17, Al 0.08, Ca 0.03, Mg 0.05, Si 0.12, Ti 0.01 [72]. With current high-throughput techniques involving multi-reactor studies, such a range of promoters can nowadays be explored more easily and in future this may lead to new and improved catalysts.

1.5 Catalyst supports

Compared to selection of the main metal, the influence of the support was considered to be rather secondary [1]. Nevertheless, the nature of the support can be of great importance and provides size, shape, and strength to the catalyst particle and, moreover, determines metal dispersion and metal distribution. As a consequence, it influences substrate reaction and diffusion. In addition, the surface properties of the support may be responsible for additional promotion and cause secondary reactions. In hydrogenation, a great many materials may be used as catalyst carrier, but mostly carbon, alumina, or silica is applied. In general, catalytic reactions require a high dispersion of the active metal that should be sufficiently accessible by selecting the appropriate porous structure. The nature of the support in combination with the catalyst preparation method determines the porosity and the metal distribution across the catalyst particle and based hereon catalysts can be classified as eggshell, uniform, or intermediate. Reaction conditions may limit the choice of support, which should be stable at the temperature used and not react with the feedstock, solvent, or reaction products.

For certain reactions, the porous structure of the support can be quite critical, especially for highly reactive molecules like acetylenes or for large molecules as triglycerides where diffusion limitation may lead to overhydrogenated products. To block alkene hydrogenation in Pd-catalyzed alkyne hydrogenation, the palladium surface must be covered with alkyne as much as possible. Therefore, most commercial acetylene hydrogenation catalysts have a very low loading of only a few hundred ppm Pd and, moreover, have an eggshell structure with a thin Pd layer of only 20–100 μm thickness. In addition, to further aid diffusion, the alumina is calcined at high temperatures to create a low surface area and wide pores [73, 74]. However, even for these eggshell catalysts with wide pores and a very thin layer of Pd, mass transfer still influences selectivity because of the high reactivity of the reaction product ethylene [75]. Other low surface area supports used especially for selectively poisoned catalysts or when a basic nature is required are calcium carbonate, barium sulfate, and magnesium oxide.

For few reactions does the success of hydrogenation depend so much on the structure and texture of the catalyst as it does in edible oil and fatty acid hydrogenation. In hydrogenation of edible oils, most, but not all, of the carbon–carbon double bonds of the triglycerides have to be hydrogenated, which elevate the melting point of the product and improve oxidative stability resulting in an improved shelf-life. The degree of hydrogenation is controlled by adapting the amount of hydrogen, reaction temperature, time, and the nature of the catalyst [27]. Narrow pores in edible oil hydrogenation hamper diffusion and lead to full hydrogenation of some molecules. Even a tiny amount of these high-melting (60°C) fully saturated triglycerides will cause "sandiness" of the fat. Thus, for edible oil hydrogenation, very wide pores having a diameter of 4–15 nm, a total absence of narrow pores <4 nm, and very small particles of only a few microns (= short pores) are essential. A very narrow particle size distribution is required to allow fast filtration of these small particles. Increasing the average pore size above 2 nm greatly improves activity, while a sharp decrease in solid fat content is observed, indicative of a decline in over-hydrogenated, high-melting triglycerides [76].

Remarkably, while a fast diffusion is required for hydrogenation of edible oils (MW 890), in contrast in the hydrogenation of the smaller but still large fatty acids (MW 280), narrow pores are required. The explanation is that during fatty acid hydrogenation, a rapid deactivation of the catalyst takes place. This deactivation is mainly due to loss of Ni surface area. In the absence of hydrogen, all Ni eventually dissolves, according to eq. (1.1).

$$Ni + 2RCO_2H \Leftrightarrow Ni(RCO_2)_2 + H_2 \tag{1.1}$$

After dosing, the catalyst/fatty acid slurry into the hydrogenation reactor, hydrogen will suppress, or even reverse, the reaction and eventually equilibrium is established in which the dissolved Ni level mainly depends on hydrogen pressure. Even at higher pressures, significant deactivation occurs via transport of nickel from the smallest (and most active) crystallites to the larger crystallites (similar to Ostwald ripening). In this process, the larger Ni soap molecule may act as a vehicle in transporting Ni. The net effect is an irreversible loss of Ni surface area. With narrow pores, the larger Ni soap molecules containing two fatty acids do not fit. As a consequence, catalysts with pores wide enough to allow rapid diffusion of the smaller fatty acids, but too narrow to accommodate Ni soaps, show an enhanced stability [76]. A similar effect of dispersion loss using metal soap as a vehicle for metal crystallite growth has been described for copper in the full hydrogenation of fatty acids to saturated fatty alcohols [77, 78]. Because of the high hydrogen pressure applied to protect the Ni and Cu catalysts against deactivation and metal dissolution, selective hydrogenation of di-enoic acids to mono-enoic acids is hardly possible.

1.6 Conclusions

Hydrogenation is an extremely versatile process with an atom economy close to 100% for several reactions. The process usually does not generate any waste, apart from sometimes water, provided that the catalyst is selective and the excess hydrogen (and solvent) can be recycled. Therefore, it is expected that catalytic hydrogenation will remain one of the major contributors toward sustainable production of chemicals. The success of current hydrogenation catalysts is based on the cooperation of three catalyst components: main active metal, promoter, and catalyst support. In addition, catalyst preparation techniques and characterization, not discussed here, have made great contributions.

Depending on selectivity, costs, filtration properties, metal recyclability, etc., either precious (PGMs) or base metals are selected. Selectivity generally is the most important criterion. Promoters do not only contribute to selectivity but may also improve catalyst activity, stability, regenerability, and processability. No sophisticated software for selecting catalyst candidates is yet available covering both base metals and PGMs for a wide range of processes. An exception is The Catalytic Reaction Guide, a commercial app designed to provide mobile access to catalyst recommendations on over 150 reactions but mostly for precious metal catalysts [79].

Most of current commercial catalysts have been optimized in the traditional way. Few ternary and quarternary metal systems have been explored in a wide concentration range, if at all. It is expected that high-throughput technology [80, 81], predictive modeling [82], and more sophisticated (operando) catalyst characterization will further facilitate the development of new and improved catalysts. Especially more complicated multicomponent systems that were not easily accessible until recently can be easily explored nowadays. In addition, modeling gradually becomes more predictive. For example, a new methanol catalyst based on Ni and Ga was discovered through a descriptor-based computational analysis by comparing the traditional Cu/Zn catalyst with thousands of other materials in the database and this catalyst has been shown experimentally to be particularly active and selective [82].

References

[1] Rylander PN. Hydrogenation methods. Best synthetic methods. London, San Diego, New York, Academic Press, 1985, 1–186.
[2] Sabatier P, Senderens JB. New synthesis of methane. C R Acad Sci Paris 1902, 134, 514–6.

[3] Normann W. Verfahren zur Umwandlung ungesättigter Fettsäuren und deren Glyzeride in gesättigte Verbindungen, German Patent 139,457, 1902.
[4] Ellis C. The hydrogenation of oils, catalyzers and catalysis. New York, D. Van Nostrand Company, 1914.
[5] Schmidt SR. Improving Raney® Catalysts through surface chemistry. Top Catal 2010, 53, 1114–20.
[6] Mittasch A, Schneider G. Producing compounds containing carbon and hydrogen. Patent US1201850, 1916, to BASF AG. See also German patents DRP 293787 (1913), DRP 295202 (1914), DRP 295203 (1914), to BASF AG.
[7] Fischer F, Tropsch H. Über die Reduktion des kohlenoxyds zu Methan am Eisenkontakt under Druck. Brennstoff-Chemie, 1923, 4, 193–7.
[8] Casci JL, Lok CM, Shannon MD. Fischer–Tropsch catalysis: The basis for an emerging industry with origins in the early 20th century. Catal Today 2009, 145, 38–44.
[9] Johnstone RAW, Wilby AH, Entwistle ID. Heterogeneous catalytic transfer hydrogenation and its relation to other methods for reduction of organic compounds. Chem Rev 1985, 85, 129–70.
[10] Horiuti J, Polanyi M. Exchange reactions of hydrogen on metallic catalysts. Trans Faraday Soc 1934, 30, 1164–72.
[11] De Vries JG, Elsevier CJ. ed. The Handbook of Homogeneous Hydrogenation. Darmstadt, WILEY-VCH Verlag GmbH, 2007.
[12] Baiker A, Blaser H-U. Enantioselective catalysts and reactions. In: Ertl G, Knözinger H, Weitkamp J, ed. Handbook of heterogeneous catalysis. New York and Weinheim, Wiley-VCH, 1997, 2422–36.
[13] Blaser H-U. Schnyder A, Steiner H, Rössler F, Baumeister P. Selective hydrogenation of functionalized hydrocarbons. In: Ertl G, Knözinger H, Schüth F, Weitkamp J, ed. Handbook of heterogeneous catalysis. New York and Weinheim, Wiley-VCH, 2008, 3284–308.
[14] Blaser H-U, Malan C, Pugin B, Spindler F, Steiner H, Studer M. Selective hydrogenation for fine chemicals: Recent trends and new developments. Adv Synth Catal 2003, 345, 103–51.
[15] Kulkarni A, Török B. Heterogeneous catalytic hydrogenations as an environmentally benign tool for organic synthesis. Curr Org Synth 2011, 8, 187–207.
[16] Nishimura S. Handbook of heterogeneous catalytic hydrogenation for organic synthesis. New York, Wiley 2001.
[17] Bartholomew CH, Farrauto RJ. Fundamentals of industrial catalytic processes, 2nd edn. New York and Weinheim, Wiley-Interscience, 2005, 487–559.
[18] Chen B, Dingerdissen U, Krauter JGE, et al. New developments in hydrogenation catalysis particularly in synthesis of fine and intermediate chemicals. Appl Catal A-Gen 2005, 280, 17–46.
[19] Arnold H, Döbert F, Gaube J. Selective hydrogenation of hydrocarbons. In: Ertl G, Knözinger H, Schüth F, Weitkamp J, ed. Handbook of heterogeneous catalysis. New York and Weinheim, Wiley-VCH, 2008, 3266–83.
[20] Freidlin LK, Kaup YY. Selectivity and stereospecificity in hydrogenation of acetylene hydrocarbons on metal catalysts. Doklady Akad. Nauk SSSR 1963, 152, 1383–6.
[21] Johnson Matthey Precious metal management, 2017 (Accessed 1 March 2017, at http://www.platinum.matthey.com/prices/price-charts).
[22] London Metal Exchange, 2017 (Accessed 1 March 2017 at https://www.lme.com).
[23] Nerozzi F. Heterogeneous catalytic hydrogenation. Platinum group metals as hydrogenation catalysts in a two-day course. Platinum Metals Rev, 2012, 56, 236–41.
[24] Bond GC. The Mechanism of the hydrogenation of unsaturated hydrocarbons on transition metal. Adv Catal 1965, 15, 91–226.

[25] Bond GC, Webb G, Wells PB, Winterbottom JM. The hydrogenation of alkadienes. Part I. The hydrogenation of buta-1,3-diene catalysed by the Noble Group VIII metals. J Chem Soc 1965, 3218–27.
[26] Rylander PN. Catalytic processes in organic conversions. In: Anderson JR, Boudart M, ed. Catalysis: Science and technology. Berlin, Springer-Verlag, 1983, 1–38.
[27] Patterson HBW. Hydrogenation of fats and oils: Theory and practice. Illinois, AOCS Press, 1994.
[28] Thakur DS, Kundu A. Catalysts for fatty alcohol production from renewable resources. J Am Oil Chem Soc 2016, 93, 1575–93.
[29] Rittmeister W. Process for the production of unsaturated fatty alcohols. Patent US3193586, 1965, to Dehydag GmbH.
[30] Dry ME. FT catalysts . In: Steynberg A, Dry ME, ed. Fischer-Tropsch technology, Volume 152, 1st edn. Amsterdam, Elsevier, 2004, 533–93.
[31] Merger F, Priester CU, Witzel T, Koppenhoefer G, Harder W. Preparation of 2,2-disubstituted pentane-1,5-diamines. Patent US5166443, 1992, to BASF AG.
[32] Ostgard DJ, Röder S, Jaeger B, Berweiler M, Finke N, Lettmann C, Sauer J. Process for the preparation of 3-aminomethyl-3,5,5-trimethylcyclohexylamine. Patent US6437186, 2002, to Degussa.
[33] Kukula P, Studer M, Blaser H-U. Chemoselective Hydrogenation of α,β -Unsaturated Nitriles. Adv Synth Catal 2004, 346, 1487–93.
[34] Hugon A, Delannoy L, Krafft JM, Louis C. Selective hydrogenation of 1,3-butadiene in the presence of an excess of alkenes over supported bimetallic gold–palladium catalysts. J Phys Chem C 2010, 114, 10823–35.
[35] Kolli NE, Delannoy L, Louis C. Bimetallic Au–Pd catalysts for selective hydrogenation of butadiene: Influence of the preparation method on catalytic properties. J Catal 2013, 297, 79–92.
[36] O'Brien CP, Dostert KH, Schauermann S, Freund H-J. Selective hydrogenation of acrolein over Pd Model catalysts: Temperature and particle-size effects. Chemistry 2016, 22, 15856–63.
[37] Englisch M, Jentys A, Lercher JA. Structure sensitivity of the hydrogenation of crotonaldehyde over Pt/SiO_2 and Pt/TiO_2. J Catal 1997, 166, 25–35.
[38] Singh UK, Vannice MA. Kinetics of liquid-phase hydrogenation reactions over supported metal catalysts – A review. Appl Catal A-Gen 2001, 213, 1–24.
[39] Pina G, Louis C, Keane MA. Nickel particle size effects in catalytic hydrogenation and hydrodechlorination: Phenolic transformations over nickel/silica. Phys Chem Chem Phys 2003, 1924–31.
[40] Horn G, Frohning CD. Supported catalysts and a process for their preparation. Patent US5155084, 1992, to Hoechst AG.
[41] Ellis PR, James D, Bishop PT, Casci, John JL, Lok CM, Kelly GJ. Synthesis of high surface area cobalt-on-alumina catalysts by modification with organic compounds. In: Davis BH, Occelli ML, ed. Advances in Fischer-Tropsch synthesis, catalysts, and catalysis. Boca Raton, United States, CRC Press, 2010, 128, 1–16.
[42] Gaube J, Klein H-F. The promoter effect of alkali in Fischer-Tropsch iron and cobalt catalysts. Appl Catal A-Gen 2008, 350, 126–32.
[43] Borninkhof F, Geus JW, Verhaak MJFM. Process for preparing primary amines and catalyst system suitable therefore. Patent US5571943A, 1996, to Engelhard De Meern BV.
[44] Deckers G, Diekhaus G, Dorsch B, Frohning CD, Horn G, Horrig HB. Hydrogenation catalyst, a process for its preparation and use thereof. Patent US5600030, 1997, to Hoechst AG.
[45] Shannon MD, Lok CM, Casci JL. Imaging promoter atoms in Fischer-Tropsch cobalt catalysts by aberration-corrected scanning transmission electron microscopy J Catal 2007, 249, 41–51.

[46] Jacobs G, Das TK, Zhang Y, Li J, Racoillet G, Davis BH. Fischer–Tropsch synthesis: Support, loading, and promoter effects on the reducibility of cobalt catalysts. Appl Catal A-Gen 2002, 233, 263–81.
[47] Jacobs G, Ma W, Davis BH. Influence of reduction promoters on stability of cobalt/g-alumina Fischer-Tropsch synthesis catalysts. Catalysts 2014, 4, 49–76.
[48] Wang J, Chernavskii PA, Khodakov AY, Wang Y. Structure and catalytic performance of alumina-supported copper–cobalt catalysts for carbon monoxide hydrogenation. J Catal 2012, 286, 51–61.
[49] Bailey S, Bonne RLC, Watson MJ, Griffiths C, Booth JS. Selective hydrogenation process and catalyst therefore. Patent WO2004108638, 2004, to Johnson Matthey Plc.
[50] Weiss AH, LeViness S, Nair V, Guczi L, Sarkany A, Schay Z, The effect of Pd dispersion in acetylene selective hydrogenation. Proceedings of the 8th International Congress of Catalysis, Vol. 5, Dechema, Berlin, 1984, 591–600.
[51] Al-Ammar AS, Webb G. Hydrogenation of acetylene over supported metal catalysts. Part 2. [14C] tracer study of deactivation phenomena. J Chem Soc Farad T 1 1978, 74, 657–64.
[52] Al-Ammar AS, Webb GJ. Hydrogenation of acetylene over supported metal catalysts. Part 3. [14C] tracer studies of the effects of added ethylene and carbon monoxide on the reaction catalysed by silica-supported palladium, rhodium and iridium. J Chem Soc Farad T 1 1979, 75, 1900–11.
[53] Rase HR. Handbook of commercial catalysts/Heterogeneous catalysts. New York, CRC Press 2000, 305.
[54] Kuiper J. Process for the selective hydrogenation of triglyceride oils with a metallic catalyst in the presence of ammonia. Patent US4278609, 1981, to Lever Brothers Corp.
[55] Kieboom APG, Van Randwijk G. Hydrogenation and hydrogenolysis in synthetic organic chemistry. Delft, Delft University Press, 1977, 141–8.
[56] Hoffer BW, Van Langeveld AD, Griffiths C, Lok CM, Moulijn JA. Enhancing the start-up of pyrolysis gasoline hydrogeneration reactors by applying tailored ex situ presulfided Ni/Al_2O_3. Fuel 2004, 8, 1–8.
[57] Hoffer BW, Devred F, Kooyman PJ, Van Langeveld AD, Bonné RLC, Griffiths C, Lok CM, Moulijn JA. Characterization of ex situ presulfided Ni/Al_2O_3 catalysts for pyrolysis gasoline hydrogenation. J Catal 2002, 209, 245–55.
[58] Goossens E, Donker R, Van den Brink,. F. Reactor runaway in pyrolysis gasoline hydrogenation. In: Froment GF, Delmon B, Grange P, ed. Studies in surface science and catalysis. Amsterdam, Elsevier 1997, 106, 245–54.
[59] Fischer L, Dubreuil A-C, Thomazeau C. Process for preparing a Ni/Sn supported catalyst for the selective hydrogenation of polyunsaturated hydrocarbons. Patent US20110166398 (2011), to IFP New Energies.
[60] Martin GA. A quantitative approach to the ensemble model of catalysis by metals. Catal Rev 1988, 30, 519–62.
[61] Rekker T, Reesink BH, Borninkhof F. Process for the production of cyclohexane. Patent US5856603, 1999, to Engelhard Corp.
[62] Gallezot P, Richard D. Selective hydrogenation of α,β-unsaturated aldehydes. Catal Rev 2006, 40, 81–126.
[63] Mäki-Arvela P, Hájek J, Salmi T, Murzin DY. Chemoselective hydrogenation of carbonyl compounds over heterogeneous catalysts. Appl Catal A-Gen 2005, 292, 1–49.
[64] Rodiansono R, Khairi Y, Hara T, Ichikunia N, Shimazu S. Highly efficient and selective hydrogenation of unsaturated carbonyl compounds using Ni–Sn alloy catalysts. Catal Sci Technol 2012, 2, 2139–45.
[65] Mahata N, Raghavan KV, Vishwanathan V. Influence of alkali promotion on phenol hydrogenation activity of palladium/alumina catalysts. Appl Catal A-General 1999, 182, 183–7.

[66] Amada Y, Shinmi Y, Koso S, Kubota T, Nakagawa Y, Tomishige K. Reaction mechanism of the glycerol hydrogenolysis to 1,3-propanediol over Ir–ReO_x/SiO_2 catalyst. Appl Catal B-Environ 2011, 105, 117–27.
[67] Nakagawa Y, Shinmi Y, Koso S, Tomishige K. Direct hydrogenolysis of glycerol into 1, 3-propanediol over rhenium-modified iridium catalyst. J Catal 2010, 272, 191–4.
[68] Nakagawa Y, Tamura M, Tomishige K. Catalytic materials for the hydrogenolysis of glycerol to 1,3-propanediol. J Mater Chem A 2014, 2, 6688–702.
[69] Lee CS, Aroua MK, Daud WMAW, Cognet P, Pérès-Lucchese Y, Fabre P-L, Reynes O, Latapie L. A review: Conversion of bioglycerol into 1,3-propanediol via biological and chemical method. Renew Sust Energ Rev 2015, 42, 963–72.
[70] Pritchard J, Filonenko GA, Van Putten R, Emiel, Hensen EJM, Pidko EA. Heterogeneous and homogeneous catalysis for the hydrogenation of carboxylic acid derivatives: History, advances and future directions. Chem Soc Rev 2015, 44, 3808–33.
[71] Sánchez MA, Torres GC, Mazzieri VA, Pieck CL. Selective hydrogenation of fatty acids and methyl esters of fatty acids to obtain fatty alcohols–a review. J Chem Technol Biotechnol 2017, 92, 27–42.
[72] Ansmann A, Benisch C, Basler P et al. Als Katalysatorvorstufe geeignete oxidische Masse. Patent DE2001151558 (2003), to BASF AG.
[73] Winterbottom, JM. Catalytic Hydrogenation and Dehydrogenation. In: Pearce R, Patterson WR, ed. Catalysis and chemical processes. London, Blackie and Sons, 1981, 12, 313–27.
[74] Bailey S, King F. Catalytic hydrogenation and dehydrogenation. In: Sheldon RA, Van Bekkum H, ed. Fine chemicals through heterogeneous catalysis. Weinheim, Wiley-VCH, 2001, 351–62.
[75] Sorbier L, Gay A-S, Fecant A, Moreaud M, Brodusch N. Measurement of palladium crust thickness on catalysts by optical microscopy and image analysis. Micros Microanal 2013, 19, 293–9.
[76] Lok CM. The 2014 Murray Raney Award Lecture: Architecture and preparation of supported nickel catalysts. Top Catal 2014, 57, 1318–24.
[77] Ladebeck J, Regula T. Fatty methyl ester hydrogenation: Application of chromium free catalysts. Stud Surf Sci Catal 1999, 121, 215–20.
[78] Ladebeck J, Regula T. Copper-based chromium free hydrogenation catalysts. In: Ford ME, ed. Catalysis of organic reactions. New York, Marcel Dekker, 2001, 403–13.
[79] The catalytic reaction guide. Accessed May 1, 2017, at http://www.jmfinechemicals.com/catalytic-reaction-guide.
[80] Ras EJ, McKay B, Rothenberg G. Understanding catalytic biomass conversion through data mining. Top Catal 2010, 53, 1202–8.
[81] Van der Waal JK, Klaus G, Smit M, Lok CM. High-throughput experimentation in syngas based research. Catal Today 2011, 171, 207–10.
[82] Studt F, Sharafutdinov I, Abild-Pedersen F, Elkjær CF, Hummelshøj JS, Dahl S, Chorkendorff I, Nørskov JK. Discovery of a Ni-Ga catalyst for carbon dioxide reduction to methanol. Nat Chem 2014, 6, 320–4.

Stephen R. Schmidt

2 The Raney® catalyst legacy in hydrogenation

2.1 History

2.1.1 Genesis

Murray Raney's 1925 invention [1], a "sponge" (or "skeletal") metal catalyst that bears his name, is not only recalled as one of the unique discoveries in applied catalysis history but survives over 90 years later in a significant commercial niche for hydrogenating organic compounds. Earlier, nickel catalysts were produced by reduction of nickel oxide [2], or were of supported type [3], and similarly many other evolved versions of nickel catalysts [4] subsequent to Raney® Ni have been supported. These employ starting materials comprised a metal precursor in soluble form, normally an acid salt. Typically, a series of steps [5] including acid–base neutralization/precipitation, thermal decomposition, and eventual reduction in hydrogen produces the active metallic nickel grains from the initial oxidized state. The Raney process described in this chapter departed from these precedents in a way that is comparatively elegant and streamlined.

The opportunity inspiring Mr. Raney's breakthrough was the assignment from his employer, Lookout Refining Co. in Chattanooga, TN, to make a better catalyst for hydrogenating cottonseed oil into shortening [6]. He succeeded using an approach resembling a hydrogen production process of his era, which selectively removed silicon from an iron–silicon alloy [7]. Raney's earliest catalyst version was based on a synthetic Ni–Si alloy, only surpassed later by the more active and more familiar Ni–Al precursor. In essence, the insolubility and corrosion resistance of nickel in the "leachant" (caustic soda) leaves it in a new metallic form.

What Raney found is shown respectively in this overall process and net chemical reaction:

$$\text{Ni–Al alloy} + \text{NaOH solution} \rightarrow \text{Ni "sponge" (with Al)} + \text{Na aluminate} + H_2 \quad (2.1)$$

$$Al + OH^- + H_2O \rightarrow AlO_2^- + 3/2H_2 \quad (2.2)$$

Raney assessed his own contribution with humility and honesty later in life: "I was just lucky ... I had an idea for a catalyst and it worked the first time," but of course creativity rarely becomes useful unless a would-be inventor knows enough to look at a relevant problem.

Later (in 1950), Raney formed his own company and remained active in its R&D work through the early 1960s, before selling the company to W.R. Grace and Co. Many

https://doi.org/10.1515/9783110545210-002

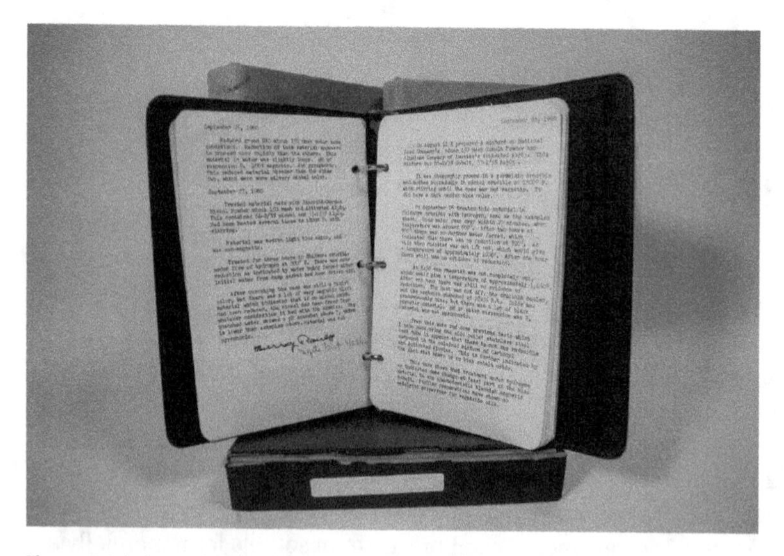

Figure 2.1: Lab notebooks of Murray Raney, ca. 1956–1965. Photo by Rebecca Huynh.

of the notebooks kept by Mr. Raney and his assistants in the 1950–1960s were transferred to Grace, retained for many years by Dr. Stewart Montgomery, and are today in the possession of this author (Figure 2.1).

2.1.2 Origin of features of various hydrogenation catalysts

The essence of any synthetic route to a heterogeneous hydrogenation catalyst is manipulation of the functional element(s) into a structure that is "well dispersed." This entails both (a) arriving at the reduced or "metallic" oxidation state that is most active and (b) extending the contained metal into greater specific surface area, which improves specific activity and lowers cost by efficient raw material usage. Raney catalysts differ from supported catalysts in the route taken.

The process steps for supported catalysts rely on (a) converting between nickel's oxidized and reduced states, with hydrogen gas or a chemical reducing agent returning a precipitated solid to active form, and (b) attaching the dispersed metal "islands" to an underlying nonmetallic support such as alumina or silica. The Raney process deviates from those steps, starting with formation of metallic alloys as catalyst precursors, and in avoiding a *separate* supporting material by using the primary metals (Ni with minor Al content) as both active surface species and support or "skeleton" (Figure 2.2).

In a more inductively reasoned view, catalysts types differ in how ingredients of a "recipe" become temporary processing aids, creating latent porosity in the catalytic metals. Supported catalyst routes generally do this by extension of solids with water. The water is first used as solvent to temporarily expand the volume occupied by the

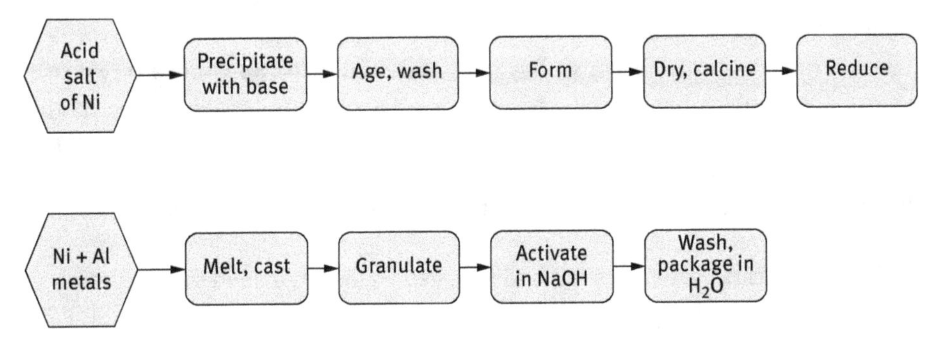

Figure 2.2: Catalyst-making process steps for precipitated (top) and Raney catalysts (bottom).

in-process metal compounds and support precursors, then as part of the two-phase mixture of precipitated solids and depleted solutions, and finally in washing or leaching solutions. Thus, water (or substituted solvent) is used as the temporary processing aid and extender, being finally removed by drying and calcining. This leaves behind the more highly dispersed, anchored, stabilized base metal structures. With use of optimized drying techniques, volume previously occupied by water at the wet in-process stage is partially retained as the pore volume of the final dried catalyst/support structure. This applies whether the support is formed prior to impregnation by metal salt solutions, or *in situ*, during precipitation steps [8] which disperse the oxidized metal-containing compounds.

In the Raney process, the role of the temporary processing aid is played by the Al metal alloyed with nickel (or other catalytic metal, e.g., cobalt or copper). In a Ni–Al alloy system formed by high-temperature melt-mixing and then cooling, the Ni is, in effect, extended and diluted to a lower density.

The precursor alloys for Raney catalysts are in the skeletal density range of 3.5–4.2 $g.mL^{-1}$ versus 8.9 $g.mL^{-1}$ in nickel's pure elemental form. A simple model based on weighted averaging of element densities predicts measured density of alloy phases reasonably well.

The Ni present in the alloy doesn't primarily exist as individual atoms in a solid solution, but as part of a mixture of crystalline binary compounds, mainly $NiAl_3$ and Ni_2Al_3 [9]. There may also be more complex ternary or quaternary (etc.) phases and structures involved when promoter elements such as Cr or Mo are added [10], some of these phases being less easily leached than binary compounds.

The substructure of Ni within the structure of each parent alloy compound is retained to a different degree upon activation by "caustic" (NaOH) solution, also depending on specific leaching methods [11]. In essence, the more Ni-rich phase Ni_2Al_3 approximately retains its Ni substructure after activation, but by contrast the less Ni-rich phase $NiAl_3$ is dissolved and then condensed, involving restructuring of Ni atoms into a somewhat imperfect face-centered cubic Ni structure. The practically achievable pore volume of the slurry catalysts from the different phases ranges from about 0.1 to 0.2 $mL.g^{-1}$. Values in this range are smaller than theoretical [12] and also appear surprisingly small because they are normalized to the weight of a very dense material (primarily Ni), but when placed on a pore volume/total volume basis, this amounts to roughly 40–60% of the particles' volume. In any case, the pore volume arises from partial retention of the space that had been occupied by Al in the Ni–Al alloy precursor, analogous to the retention of water volume as final pore volume in a supported catalyst.

The Raney catalyst also contains Al species included as a solid solution component (within nickel) and a second hydrated alumina phase scattered within the Ni grain structure [13]. Many properties of Raney catalysts, including surface area and surface chemistry, are affected significantly by the amounts of aluminum and alumina that are present [11].

2.1.3 Taxonomy of hydrogenation catalysts

Figure 2.3 summarizes how a number of the common catalyst types relate to each other, restricting to hydrogenation as a primary class of applications.

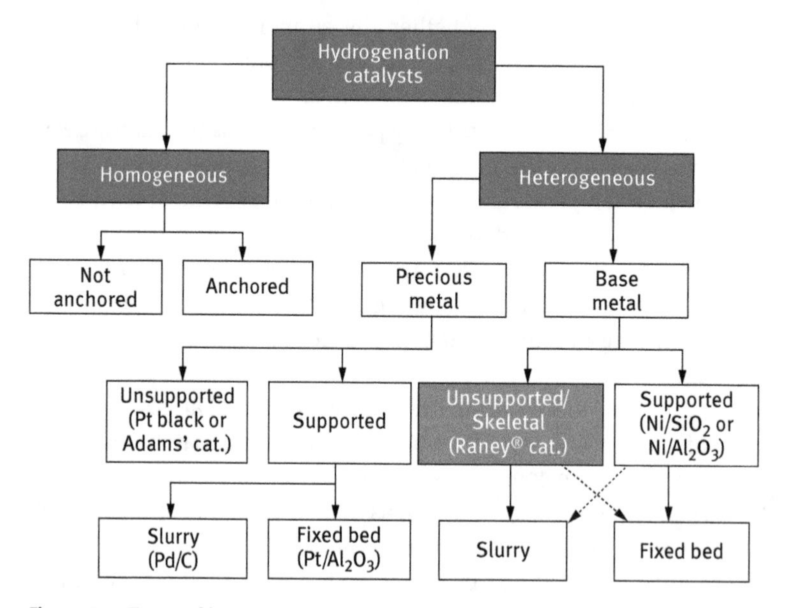

Figure 2.3: Types of hydrogenation catalysts (examples in parentheses).

In the figure, the catalyst types are exemplified by specific versions in parentheses. Raney catalysts are unsupported base metal type, both in slurry form for stirred or bubble column reactors and fixed-bed form for packed tube reactors.

2.1.4 Essential features of Raney process and resulting catalysts

A simplified overall plant scale manufacturing process is outlined in Figure 2.4.

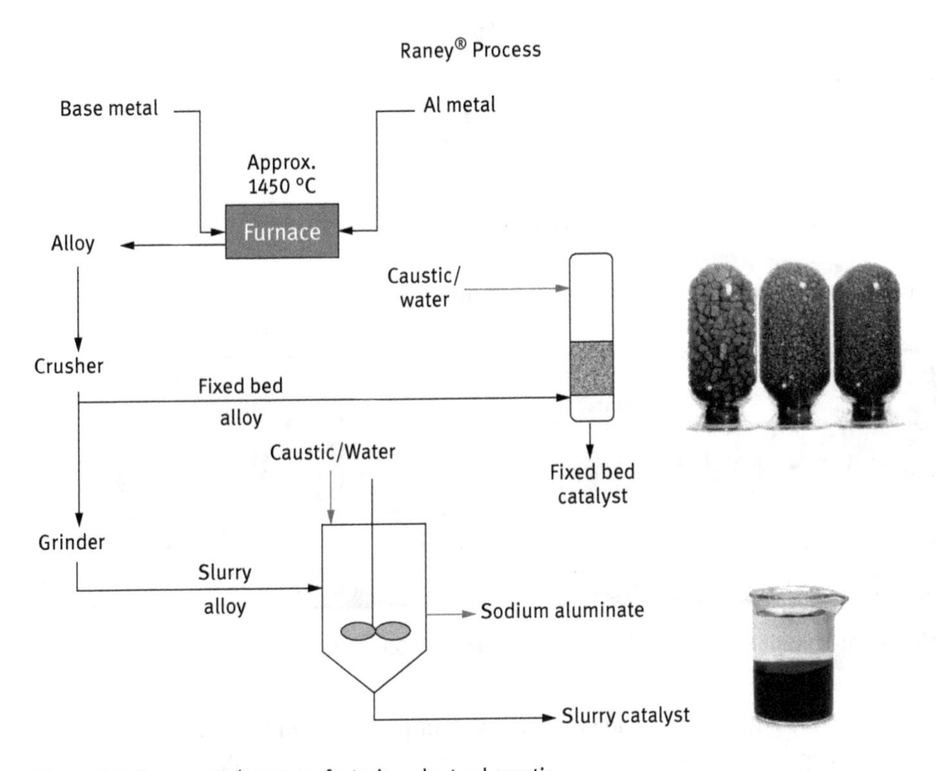

Figure 2.4: Raney catalyst manufacturing plant schematic.

The niche occupied by Raney catalysts evolved partly from physical and chemical properties that distinguish it from other types. Some common chemical and physical property ranges are summarized in Table 2.1 for slurry catalysts and Table 2.2 for fixed-bed catalysts.

Table 2.1: Property ranges for Grace Slurry Raney® catalysts.

Catalyst type	Al%	Median diam. (μM)	BET S.A. ($m^2.g^{-1}$)	Active S.A. ($m^2.g^{-1}$)
Unpromoted Ni	4–6	25–65	60–80	40–45
Fe–Cr-promoted Ni	7–12	25–55	120–140	40–50
Mo-promoted Ni	4–7	25–55	70–80	30–35
Promoted cobalt	4–5	25–55	70–80	25–30
Copper	2–5	30–65	15–40	NA

Table 2.2: Fixed-bed Raney Ni physical features.

Property	Range
Mesh size	8–12 to 3–6
Diameter (mm)	2–3 to 4–8
Particle density ($g.mL^{-1}$)	3.5–3.6
Packed density ($kg.L^{-1}$)	1.7–1.9
Crush strength (lbs)	15–115
BET surface area ($m^2.g^{-1}$)	15–35
Ni content (wt%)	47–56
W.R. Grace Products	

Slurry alloy-to-catalyst genesis is pictured in Figure 2.5.

Figure 2.5: Activation of powdered Ni–Al alloy added to NaOH solution. Photo by Edward Laughlin.

The deceptive simplicity of the slurry activation process is indicated in these three pictured steps. Washing steps ensue after the stage pictured at right.

Alloy precursors for the fixed-bed Raney catalysts are shown in Figure 2.6.

Figure 2.6: Alloy precursors for fixed-bed Raney catalysts.

Compared to slurry catalysts, fixed-bed catalysts derived from conventional granular-type alloys shown above are leached to a much lesser degree, both in terms of required NaOH concentration and fraction of aluminum removed. Alternative approaches to fixed-bed catalysts are revisited briefly in Section 2.4.3.

The high density of Raney slurry catalysts, at about 6–7 $g.mL^{-1}$ particle density (the effective specific gravity including hydrogen effects), leads to rapid settling in liquid systems, when compared to less dense carbon- or silica-supported catalysts of comparable particle size.

High metal content leads to high strength and rapid heat transfer for the fixed-bed Raney versions. The high density is not as advantageous as in the case of the fluidized reactors, where settling could be a rate-limiting processing step.

2.2 Applications

2.2.1 Road map of utility of Raney catalysts

The reducible functional groups shown overlap significantly with those of other established hydrogenation catalysts, e.g., supported Ni or PM/carbon, which are applied in many similar instances. Thus, the broad field of chemical hydrogenations is well represented by known Raney applications. Making the optimum choice of a catalyst for a given commercial involves distinctions about cost, activity, selectivity, and stability, which in turn may require customization.

A brief synopsis of some of these key applications in which Raney catalyst is highly effective, in clockwise order of the routes, is shown in Figure 2.7 (starting from lower left).

2.2.2 Sorbitol

Sorbitol is an important food additive and pharma intermediate, produced by batch reduction of the cornstarch-derived sugar D-glucose ("dextrose"). The monomer largely exists in the cyclic "pyranose" isomer, by internal hemiacetal formation. Variously used feedstocks differ in monomeric dextrose purity, depending on upstream removal of the oligomers, polysaccharide chains with 2–4 glucose monomer units (e.g., maltose for $N = 2$). Even after ring opening of terminal units and reduction of the aldehyde function, the larger oligomerized molecules generally persist in product mixtures. This means that tolerance for the non-sorbitol polyol components (e.g., maltitol) in the downstream application largely constrains what feed enters the hydrogenation step. The preferred catalyst is molybdenum-promoted Raney nickel [14]. Mannitol and gluconic acid normally are the only significant catalytic by-products. Use of buffering to optimal pH, along with adequate agitation for good hydrogen transport, allows for extensive catalyst reuse and long lifetimes.

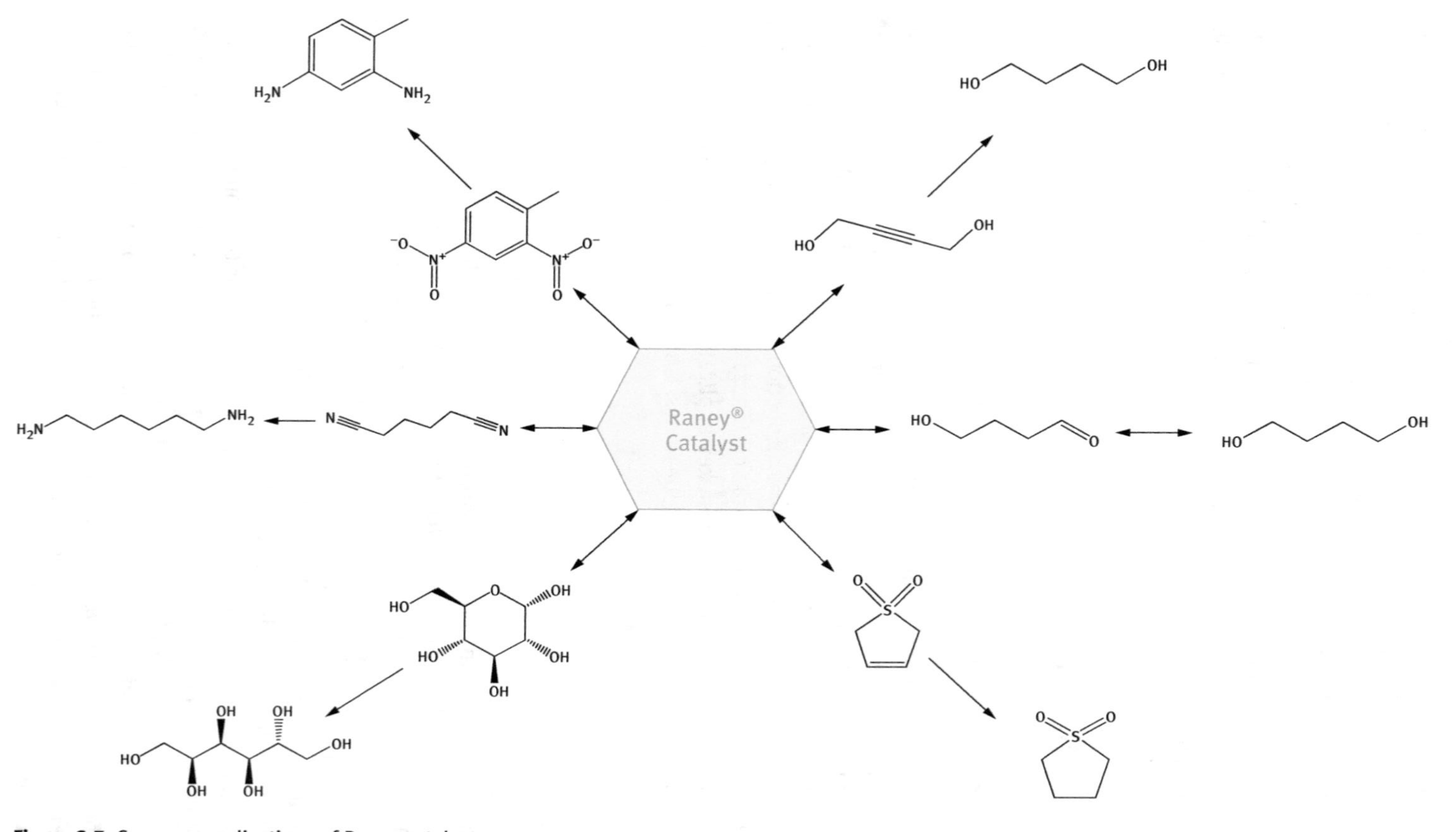

Figure 2.7: Common applications of Raney catalysts.

2.2.3 Hexamethylene diamine

Hexamethylene diamine (variously known as HMD or HMDA) is one of two reactants in the nylon 6,6 process, the other being adipic acid. In one process variant, the diamine is produced by hydrogenation of adiponitrile, the corresponding dinitrile, using a fine-particle Raney Ni which is doubly promoted (Fe and Cr are added to the alloy stage). The reactor is a "bubble column," i.e., fluidized upflow type [15] for which catalyst size distribution is optimized. Alkalis such as sodium hydroxide are added for extending catalyst life, ease of separations, and for selectivity control, including minimization of higher order amines. Practical considerations include avoidance of nitrile feed in premature contact with the catalyst (in absence of applied hydrogen) since the nitrile function is a strong adsorber on Ni surfaces.

2.2.4 Toluene diamine

Toluene diamine is produced mainly in its 2,4 and 2,6 isomers from the corresponding dinitrotoluenes. One mole of this highly reactive feedstock requires 6 mol of hydrogen for the highly exothermic reduction, one which is potentially quite damaging to the suspended Raney nickel catalyst, due to the oxidizing nature of nitrobodies. Dilute reaction mixtures are solvated by either simple alcohols [16] or water–diamine product [17] moving through multistaged upflow reactor chains. Isocyanates are formed downstream and reacted with polyols to make polyurethanes. The latter are finally consumed in foam and fiber applications, for automotive components and clothing.

2.2.5 1,4 Butanediol

2.2.5.1 Reppe process

This diol product is used as an intermediate to tetrahydrofuran and gamma butyrolactone with eventual outlets to polyesters, polyurethanes, and other polymers. In the Reppe process, the feedstock is 1,4-butynediol formed by hydroformylation of acetylene. The catalyst is a fixed-bed Raney nickel formed in the hydrogenation reactor itself by partial leaching of large-granule Ni–Al alloy. Butanol is the main by-product. Its level increases over time and is monitored as one of the markers of gradual catalyst decay [18].

2.2.5.2 Allyl alcohol process

This alternate version of the process starts with the hydroxyaldehyde formed by reacting allyl alcohol with CO and H_2. Both slurry [19] and fixed-bed [20] processes are practiced industrially; some instances of each use Raney nickel catalyst with promoters. Selectivity challenges are somewhat similar to those of the Reppe process, with *n*-butanol being formed through several possible routes involving dehydration.

2.2.6 Other Applications

Other significant industrial products traditionally based on Raney catalysts include sulfolane, a solvent formed from the starting material sulfolene [21], plus (not shown in Figure 2.7) amino acids from amino alcohols, notably as used for glyphosate intermediate [22], and miscellaneous amines from corresponding nitriles.

2.3 Material properties of the catalysts

2.3.1 Importance of alloys

The phase diagram of the Ni–Al system is shown in Figure 2.8 [23].

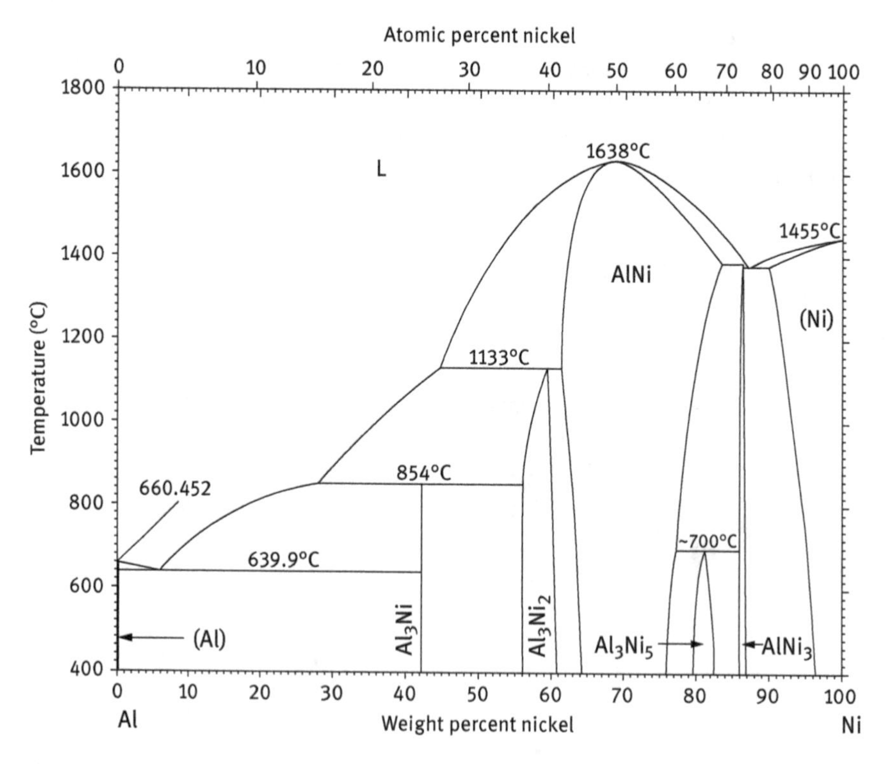

Figure 2.8: T–X phase diagram for Ni–Al alloy system.

This system contains the two previously mentioned key phases, the "named" crystalline compounds $NiAl_3$ and Ni_2Al_3. $NiAl_3$ is a "line compound" in phase diagram parlance, meaning that it has an exact composition at the 3Al:1Ni atomic ratio (42% Al by weight) with no significant "phase width." Adding Ni or Al to this composition, changing it away from the exact ratio, will finally yield a two-phase mixture at equilibrium. By contrast, the Ni_2Al_3 has significant phase width; within the range shown from ~56% to 60% Ni, the Ni or Al can be increased without forming a separate second phase. This has implications for "capacity" of a few percent of other elements (beyond Ni and Al, e.g., Cr or Mo) being added as promoters that can dissolve into the structure of Ni_2Al_3 [24].

Both compounds can be formed peritectically, meaning that a given molten composition in the vicinity of the named compound can solidify in stages as temperature drops. This initially yields a solid compound in contact with a new liquid composition (indicated by curved boundary lines in the diagram) but finally forms a two-phase solid mixture when all liquid crystallizes. With normal, "moderately fast" cooling (e.g., when a melt is poured onto a cooling slab), this quenched mixture is normally not yet at equilibrium; due to the peritectic sequence, there is more of a higher melting solid, trapped in a metastable state, than will exist at equilibrium. At less than 42% Ni (left of line compound), the phases in a solid mixture will be $NiAl_3$ and the solid solution of Ni in Al. To the right of the line, or >42% Ni, Ni_2Al_3 is in the mixture along with $NiAl_3$.

Equilibrium in the solid mixture can be achieved by annealing, i.e., heating the solid at a temperature below its melting point to rearrange metal atoms by diffusion. In effect, a pure $NiAl_3$ phase could be achieved by annealing the exact 42 wt% Ni composition, while the *ratio* of Ni_2Al_3 to $NiAl_3$ is shifted by annealing when the overall composition is in the two-phase region between 42% Ni and 56% Ni. Alternatively, annealing may equilibrate the single phase within the ~56–60% Ni overall composition range of Ni_2Al_3.

The two key alloy phases differ in a number of ways described previously [11]. At fixed conditions, the Al rich 3:1 phase reacts faster in NaOH solution ("leaches") to deplete its Al more completely over a fixed time, to thus yield a lower Al catalyst product. Conversely, the higher Ni 3:2 compound leaches more slowly to retain more Al.

Further complicating this picture, there are different mechanisms involved in the caustic leaching of each phase to form a catalyst [12]. The more Al-rich $NiAl_3$ is said to dissolve completely and then quickly recrystallize the Ni atoms to a more condensed face-centered cubic structure resembling Ni metal, albeit with differences from pure Ni in having imperfect crystallinity and some inclusion of Al species. The Al content varies, centered around 5 wt%, highly dependent on severity of leaching conditions.

By contrast, the Ni substructure of the Ni_2Al_3 phase is retained partly intact during Al removal; some chains containing Ni–Ni bonds survive the leaching rather than being completely dissolved by the leachant. This results in retention of the original specific volume (inverse of density) of the alloy and thus greater persistence of the alloy particle's size and shape in the finished catalyst, essentially less collapse than from $NiAl_3$.

The impact of making a catalyst from a given alloy phase composition (whether single- or multi-phase) has been studied repeatedly without yielding any simple answer about what is "best." There are instances of effective catalysts made from a wide range of alloy compositions. The need for various catalyst features (especially chemoselectivity, cost, and stability) has led catalyst makers to customize, combining Ni/Al variations + promoters + leaching techniques into multiple products and processes, some proprietary.

2.3.2 Other properties of the catalyst

At a more fundamental level, Fouilloux's 1983 review [11] contributed some key characterizations of the Raney catalyst as a material, e.g., connecting magnetization with Al content and also relating method of catalyst preparation to resulting properties such as structural defects, residual aluminum, and alumina. One clear consensus from numerous studies cited and extended by Fouilloux: all Raney catalysts contain some residue of the alloyed aluminum, residing in both metallic and hydrous oxide species (2–8% overall in slurry catalysts; much higher amounts for some fixed-bed types). There can be wide variation in the oxides as a subset of the total. In turn, many attributes of the catalyst derive as much from these Al species as from nickel *per se*.

Presence of hydrogen as part of the catalyst composition at its "birth" is another distinguishing feature. This hydrogen, accompanying zerovalent Ni, makes the catalyst generally ready for use, without prereduction by hydrogen in the reactor and thus minimizing the "break in" or conditioning period. It also means the surface is potentially reactive with oxygen, so drying in air is to be avoided to prevent igniting organic matter, or at a minimum, deactivating the catalyst by oxidation. This requires storage under a liquid, usually water but alternatively some organic solvent or oil. High density of particles makes this immersion in liquid quite easy for Raney catalysts.

The need for protection from air is true of *all active* hydrogenation catalysts, varying mainly in *when* this becomes necessary during their respective life cycles. The trade-off of air stability for convenience (readiness for use) is also not unique to Raney catalysts. Supported Ni catalysts can be obtained either in variously oxidized forms that require *in situ* reduction in hydrogen at time of use, or pre-reduced and encapsulated, e.g., in wax, to seal them from air. After any of these nickel catalysts is introduced to a catalytic reactor and achieves its active form, it is thereafter protected under hydrogen pressure.

Retaining even a small fraction of the copious amounts produced in alloy leaching, the Raney catalyst can contain both bubbles of H_2 visible in a macro view, as well as chemisorbed H_2 "seen" by surface characterizations. A third type of hydrogen "storage" in the form of subsurface solution or hydrides in the metal structure has been postulated, and attempts were made to quantify it [25]. In instances where heating was applied to desorb the contained hydrogen quantitatively, it is likely that some of this "extra" hydrogen actually came from the artifact of Al metal reacting with residual water. (The next section elaborates further.)

2.3.3 Surface characterizations

The most frequent attempts to connect essential material properties of the Raney catalyst with its performance have focused on the abovementioned hydrogen content and/or nickel surface area [26, 27]. Our studies in this area [28] concluded that attempts to measure chemisorbed hydrogen, whether by using conventional "adsorption mode" or thermal desorption, were likely to be inaccurate. Incomplete removal of the hydrogen originally present on the catalyst leads to undercounting of hydrogen-adsorbing sites in the subsequent adsorption experiments, while the Al + water artifact plagued the thermal desorption experiments by an (unquantifiable) false contribution. We ultimately evolved to CO as an alternative probe molecule. While use of the CO probe in itself was not novel [29], uncertainty persisted around the stoichiometry on Ni for a given catalyst, i.e., whether it involved probe molecules adsorbed directly one-on-one to a metal atom or bridged between two or more sites [30]. We hoped to determine this adsorbed CO/Ni° ratio in some new way. A "titration" method that could directly quantify active metal surface areas might correlate better with models of the surface, and with activity tests.

We eventually succeeded by using CO (introduced at ~1 atm total pressure; pulse mode in flowing helium) to displace hydrogen that had been chemisorbed onto Raney® catalysts in a prior step. This was possible because of reduced Ni's stronger affinity for CO than for hydrogen in the temperature ranges employed (adsorbing at, e.g., 0°C *via* ice water bath). We thereby "calibrated" the CO probe molecule for its stoichiometry of adsorption on nickel and cobalt. Calculation using known 1:1 stoichiometry for H radicals on surface Ni atoms [31] gave a conversion factor for "active" (chemisorbing) surface area of 1.8 $m^2.mL^{-1}$ of CO.

We were able to further validate the CO–H_2 method as measuring chemisorbed hydrogen, through strong agreement with values from thermal desorption. This entailed separating the Al–water reaction artifact from the total H_2 released.

Finally, the surface spectroscopic method X-ray photoelectron spectroscopy (XPS) was used to create a simple model predicting H_2 and CO adsorption capacities, based on relative surface populations of elements and their oxidation states (Figure 2.9).

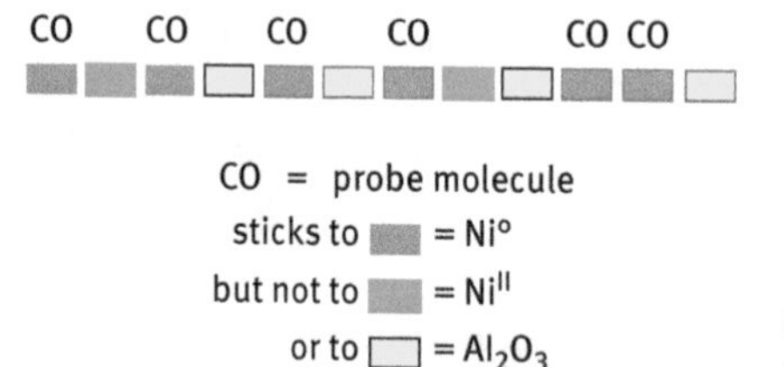

Figure 2.9: Simple model of chemisorption of CO on Raney nickel.

XPS yielded fractions of the surface occupied by each type of atomic species. Combined with assumed surface area per atom type, this predicts values for surface Ni° in variously treated samples. In turn, this was reasonably accurate in predicting the measured chemisorption values at the assumed 1:1 stoichiometry as shown in Table 2.3.

Table 2.3: Surface measurements on Raney® Ni:CO chemisorption and XPS.

Pretreatment	BET SA ($m^2.g^{-1}$)	Adsorbed CO ($mL.g^{-1}$, 0°C)	Displaced H_2 ($mL.g^{-1}$, 0°C)	Predicted H_2 (by XPS)	CO:H (CO:Ni°)
None	74	24	14	12	0.86
Stir 8 h (air)	57	16	9.2	7.0	0.88
TD,[a] 350°C/10′	59	NA	5.9[b]	NA	NA

[a] Thermal desorption of hydrogen into flowing nitrogen.
[b] *cf.* 5.3 from a second thermal desorption of same sample.

Besides improving characterization tools for Raney surfaces, the CO–H_2 displacement method showed that the percentage of total nickel atoms that are useful chemisorbing surface atoms, i.e., "dispersion," is only about 8–10%. This is an explainable, inherent feature of these catalysts, deriving from how surface area is achieved. Because the Raney process omits a separate pre-formed carrier phase ("support"), the Ni or other catalytic base metal provides both underlying structural strength and the active sites, meaning that much of the nickel is buried rather than at the catalytically useful surface. This somewhat inefficient metal usage in a Raney catalyst (cf. >20% for some supported Ni catalysts) [30] is usually accepted as a "given" and is regarded as offset by an elegant manufacturing process, to yield a very affordable catalyst. Alternatively, the modest dispersion is a challenge to be addressed, a motive for raw material substitution to further lower manufacturing cost. An unusual example of this is described in Section 2.4.2.

XPS is used in some specific examples in our work and also merits a separate introduction here among the essential surface methods. The method, also known as electron spectroscopy for chemical analysis, impinges X-rays of known energy onto a material which then emits electrons (photoelectric effect). The measured kinetic energy of emitted electrons yields info about their prior binding energy. This in turn yields info on the oxidation state(s) of an emitting element, and possible identity of compounds containing said element. The XPS technique is surface sensitive to a depth of about 50 Å, limited to the maximum "escape depth" of the emitted electrons. That makes the chemical information relevant for characterizing catalyst particle surfaces, defined in terms of their top several layers of atoms. A penetration to 50 Å actually "sees" more deeply than just those exposed atoms involved in chemisorption and defining dispersion, but this doesn't probe the particle's entire internal structure. Spherical particles of radius ranging from (e.g.) 5–25 μm have about 1,000–5,000 times greater path to their centers than what is accessed by XPS. This means XPS is only (spectroscopically) "peeling the outermost skin" of these particles to quantify various types of Ni and Al sites, indirectly informing about radial distribution of impregnated metals (e.g., "egg-shell" vs. "uniform" placements). Combined with a "bulk" chemical analysis technique such as ICP, XPS reveals how much an element is "surface enhanced," i.e., concentrated near the periphery as compared to its overall abundance in the sample.

2.4 Evolution of Raney catalysts

2.4.1 Promoters

Conventional base metal promoters can augment a Raney catalyst's features in any of several ways: increased surface area, improved chemoselectivity and activity for specific functional groups [32], greater stability against oxidation or other chemical attacks during use, etc. The promoters are implicitly defined as minority components in the material; these base metal additives typically range from 1 to 5 wt% of the overall chemical composition. They are introduced to the catalyst-making process in one of two ways, namely (1) during alloying or (2) during/after the activation step, via solution phase (being "plated" or "impregnated"). Common examples of alloyed promoters are Cr, Fe, Mo, and Cu; typical surface-applied promoters are Mo and Cr. The surface impregnation method sometimes functions more efficiently than does alloying, avoiding any promoter metal potentially being "buried" in the non-surface or "bulk" atomic layers.

Whether introduced during alloying or solution-applied after Al is leached away, a promoter alters surface chemistry of the catalyst from a baseline of standard Ni and Al species. The added sites may modify acid–base character, affecting selectivity.

In a simple view of a non-promoted hydrogenation catalyst, an ensemble of neighboring Ni sites adsorbs both the organic functional group and the hydrogen, to put the reactants in close proximity. A hydrogen molecule is said to be "activated," or to stretch and split, seemingly behaving as two independent H radicals [33], and eventually form (e.g.) two single C–H bonds from a double bond such as –C=C–.

A surface promoter may create a new type of site as an option for adsorbing organic functions on the surface. When the bond to be hydrogenated is "hetero" (carbon bonded to non-C such as -N or -O), its polarity figures in this two-site mechanism. Selective adsorption of hetero-organic functions onto surface promoter sites consisting of "hetero" pairs of metal and nonmetal atoms M–X may be favored, as both are polar. In this alternative picture of an ensemble, the neighboring Ni–Ni and M–X sites, respectively, adsorb and activate the H_2 and polar functional group (e.g., –C=O).

Both alloy leaching and post-activation promotion generally lead to oxidized, hydrolyzed promoter species such that a promoter "metal" like Mo or Cr is actually retained in a hydrous oxide form on the surface [24]; thus, the generic surface site M–X may be, e.g., Mo–OH, Cr–OH, or their conjugate base anionic forms.

Precious metal promoters (a.k.a. "dopants") can surpass what base metal promoters do, both conceptually and catalytically. The earlier use of the term "promoter" for a minority component (at ~1–5%) also implies its secondary role behind the majority component nickel (or cobalt, etc.). The element Ni, at ~95 wt% level in a Raney nickel (slurry type), serves in the primary "catalyst" role, augmented by promoters not active enough to stand alone as hydrogenation catalysts. The oxide forms of Cr, Mo, etc. cannot chemisorb hydrogen and polar organic functional groups at normal T–P conditions, preventing independent roles as hydrogenation sites.

By contrast, a precious metal like Pd or Pt dispersed onto a base metal catalyst is "noble," i.e., zerovalent; it can chemisorb hydrogen as well as organic functions. Precious metals serve independently as catalysts, at *high enough metal loadings and appropriate operating conditions* (key qualifiers revisited below). This suggests Pd or Pt could also function as an efficient cocatalyst with Ni. At the least, hydrogen spillover from the Pd or Pt to Ni seems feasible.

In the "taxonomy" presented in Figure 2.3, hydrogenation catalysts are divided into base metals versus precious metals (hereafter "PM"), but in practice, there is also a large overlap in the areas of utility for these two main types. Over a wide range of hydrogenations, these catalyst types may substitute for each other, although sometimes at modified T, P conditions or catalyst loading. Verifying this at lab stage may not yet consider catalyst cost or practicality of scale-up.

To eventually be used in manufacturing, a catalyst also needs some minimum productivity/cost ratio, and must perform under equipment constraints, such as upper pressure limits on reactors. The productivity/cost factor is one central issue in choosing between Raney catalysts and supported PM. The PM catalysts are usually both more expensive and more active at milder conditions. If an application demands high activity but there is also a "low ceiling" in achievable hydrogen pressure, the precious metals' inherently faster "activation" of hydrogen may be preferred. Their high purchase price suggests that, for commodity organic chemical production, PM catalysts still need adequate lifetimes and efficient metal reclaiming to have lowest cost overall. When this is not so, a base metal catalyst of adequate selectivity is still a competitive option.

Acknowledging that Raney and PM catalysts have different productivity/cost "balance points" and had thereby evolved to different niches, we aimed to modify one in the direction of the other. Specifically, we wanted to enhance activity and durability of a Raney catalyst toward those of PM catalysts, but at much lower cost. We anticipated that this could also expand the range of viable operating conditions, e.g., toward lower pressure.

The approach taken to enhanced activity was to deposit a minimum of precious metal ions from solution onto an existing Raney nickel surface, seeking synergistic PM–Ni interaction, i.e., beyond "promoter" effects. Electrochemical ranking predicts that a precious metal ion such as Pd^{2+} or Pt^{2+} will spontaneously be reduced to active metallic form by either a zerovalent nickel surface or its adsorbed hydrogen. In principle, this obviates any additional (deliberate) reduction step for the PM.

Literature review uncovered how this simple idea was pursued as far back as the 1930s, in the early days of Raney technology [34]. The earliest attempts used acid salts (e.g., Pt chlorides). Because a Raney catalyst's aqueous environment is naturally alkaline, this risks acid–base precipitation, possibly even as separate particles of oxides of Pd or Pt, rather than the intended redox process that would attach PM in zerovalent form onto Ni. Some prototype Pt/Ni catalysts [35] showed inconsistent activity, attributed to unacceptably short shelf-life, requiring *in situ* preparation at time of use.

We believed that an optimized PM/Ni surface promotion method would increase PM dispersion, and more consistently yield a stable, highly active catalyst. The key in achieving this was use of alkaline Pd or Pt salts, specifically ammine complex type. We distinguished the resulting new catalysts from older acid-salt prototypes in two ways: (1) XPS analysis showing different radial distribution of Pd, implying its greater dispersion, and (2) catalytic testing. A hydrogenation reaction that probes vulnerability of a base metal catalyst was chosen for the test case: hydrogenation of *p*-nitrotoluene. This compound generically represents nitroaromatic reductions (which includes, e.g., the industrially important dinitrotoluene-to-TDA processes described above). Nitro compound processes are inherently problematic for catalyst durability, because the feedstock itself is a chemical oxidizer of the catalyst metal surface. That in turn places high demand on surface hydrogen concentration, or "hydrogen availability," which could be addressed by improving the speed of hydrogen activation on metal sites.

The evidence from the *p*-nitrotoluene work (Table 2.4 and Figure 2.10) is that Pd or Pt is extremely efficient in boosting the activity of Raney Ni when applied in the improved ammine–salt method.

What amounts to only 1 PM atom in about 400–600 atoms in the overall composition increases the activity by a factor of 1.7–2.5, depending on time-on-stream. The goal of true "cocatalyst" behavior rather than mere "promotion" seems to have been achieved, probably taking the form of hydrogen spillover from PM onto a reduced Ni surface, the latter providing the majority of the effective sites. In a comparative experiment, the same amount of Pd on an alumina support "died" immediately in this testing, suggesting Ni is needed to take up and use the H_2 activated by the Pd or Pt.

Table 2.4: Activity (mmol $H_2.min^{-1}.g^{-1}$), batch-recycle *p*-nitrotoluene, 140°C/400 psig/1.86% loading.

Cycle	Basic 0.25% Pd/Ni	Acidic 0.25% Pd/Ni	Raney Ni
1	148	114	85
2	137	77	60
3	77	45	40
4	57	48	35

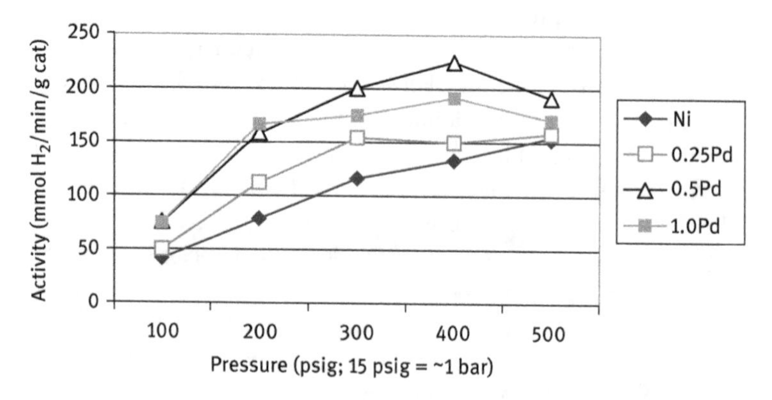

Figure 2.10: Activity of Pd-promoted and standard Raney® Ni in *p*-nitrotoluene conversion at 125°C.

Unaccompanied by Ni, the minor amount of PM atoms provides too few sites and they are overwhelmed (deactivated) by the nitro group oxidants.

The alkaline–salt method also improves on the older acid–salt method with higher activity at any equivalent PM dosing, and the radial placements of PM within the Ni pore structure also differ [36]. The periphery contains more PM when the acid–salt method is used, confirming the premise that acid–base "shock" would lead to more abrupt deposition. Finally, the goal of reducing the required H_2 pressure was also achieved for the 0.5% Pd% loading/14 bar (200 psig) combination, which achieves similar activity to Ni alone at 34 bar (500 psig). It is in fact possible to move the behavior of a base metal catalyst in the direction of a precious metal catalyst using an efficient combination of the two metal types.

2.4.2 Electroless plating

The relatively inefficient usage of nickel (or cobalt) in Raney® catalysts, a dispersion value under 10%, was noted and explained above. Beyond the scientific or academic significance of this, there has been a perpetual examination of catalyst manufacturing costs, which is affected by both Ni usage rates and raw material price. As nickel price climbs, the scrutiny of raw material costs also increases, and occasionally lower cost alternatives are proposed and tested. One approach to a cheaper catalyst is partial substitution of cheaper metals into the precursor alloy, e.g., replacing ~20–50 wt% of the main catalytic metal. (For perspective, the range for added promoters such as Cr and Mo is much lower, 1–5 wt%.) A DuPont invention based on the substitution approach [37] actually altered the alloy for Raney cobalt, replacing some of it with iron. This did not succeed in matching performance of conventional Raney cobalt or displace it from typical applications. A "diluent" effect from the new metal occurs if its inherent activity is lower than what it replaces.

Other requirements of the cheaper, substituted metal include (1) resistance to dissolving or corroding in NaOH solution and (2) attainment of high surface area via selective leaching of Al. We expected that, even after having met these goals, we could still find the surface concentration of nickel, and possibly the potential catalytic activity, to be proportionally lowered by the metal replacement. If we (e.g.) replace half of the alloyed nickel atoms, and there is no "surface enhancement" of Ni (selective concentrating at the interface), half of the resulting surface would be less active than with nickel alone. The corresponding effect of Fe in place of Co accounts for the DuPont outcome above.

The flawed prototype method of simple alloyed substitution evolved to an improved, two-step metal incorporation. This still entailed significant replacement of alloyed Ni by the diluent metal in a first step, but the partly non-nickel surface thus formed was then treated in a second step, to coat non-nickel sites. This produced higher active site concentration on the surface, aiming to match that of Raney® nickel itself. The proposed coating step would use soluble (cationic) Ni species and thus require a redox mechanism to form reduced Ni^{o} sites. The reduced metal sponge, spontaneously formed in the Raney® process, would be useful in this redox plating process. If successful, this would later avoid reduction by high-temperature hydrogen gas and retain the expected "ready for use" advantage cited earlier.

Metal deposition or "plating" in order to radically change the primary surface composition of a base metal catalyst had not been reported prior to the late 1990s. More broadly, plating of various metal objects had been well established. There are two distinct electrochemically driven processes in use; either will reductively deposit metals from solution. One version, displacement plating, uses a sacrificed portion of the metal substrate surface as the reductant. Electroless plating, by contrast, adds a reducing agent, a compound easily dehydrogenated on the substrate surface, e.g., $NaBH_4$ or NaH_2PO_2 [38]. This generates hydrogen, the intermediate which gets oxidized instead of the metal substrate. The two plating types were implemented in patented catalyst-making processes by Monsanto [39] and W.R. Grace [40], respectively.

Monsanto's plating process displaces the more expensive Ni from a catalyst surface using less expensive Cu, yet overall, the use of Ni as substrate actually *increases* metal cost because the previously used catalyst was essentially "all Cu" (Raney Cu).

The Raney Cu's low metal cost starting point accommodates this substitution of Ni substrate as long as performance is adequately improved. The chosen type of plating also affects the size of cost increase. Displacement plating substitutes Cu for Ni in one-for-one fashion (yielding dissolved Ni by-products) and thus not all of the metals employed between initial substrate and plating contribute to the final yield. In equation form, the weight change factor over starting substrate is $1 - x + y$, with 1 as the original substrate weight, x being the displaced (lost) substrate metal, and y the deposited metal. At equal molar substitution and nearly equal atomic weights for Cu and Ni, essentially $x = y$. This limits final catalyst yield to about 1, i.e., no net gain or loss, which may succeed economically depending on the extent of lifetime enhancement. The improved stability and surface area of plated Cu/Ni significantly extended catalyst life beyond Raney copper in making amino acids from amino alcohols.

The electroless plating process patented by Grace started at a different point and moved toward a different goal: high initial metal cost was to decrease, while matching existing performance. This suggested a route with a higher yield, i.e., the weight change factor $1 - x + y$ needed $y \gg x$ to approach a net $1 + y$. Further, even if added (plated) y is all Ni, the Ni content of the original weight "1" is lowered to $\ll 1$ by alloy substitution, aiming at overall Ni content much lower than typical ~95% for Raney® nickel.

Iron is the lower cost metal substituted for Ni in this second approach. Raney® iron existed prior to this invention [41] but was known to have low activity and few applications. The use as a substrate for plated Ni was expected to be unhindered by low activity as long as adequate surface area could be achieved. The other key steps identified in making plating Ni onto Raney Fe were as follows:

1. use of acid to "clean" the Raney® Fe surface (remove oxides) *in situ* to expose metal
2. inclusion of some Ni alloyed into the Fe substrate to serve as initiator "seeds" underneath and improve subsequent plating
3. use of abovementioned electroless approach with a reducing agent such as NaH_2PO_2 to achieve the improved $1 + y$ yield

This approach yielded a prototype catalyst which matched activity of Raney Mo/Ni in batch hydrogenation of dextrose to sorbitol, but with a Ni content of approximately half of the traditional catalyst. Analysis of the surface of the plated Ni/Fe catalyst by SEM, nitrogen BET, and XPS verified, respectively, that the new material was changed in surface morphology, increased in total surface area, and in fact was coated with nickel atop iron, through the electroless process.

2.4.3 Fixed bed Raney catalysts

As described earlier (Figure 2.4 and Table 2.2), the Raney alloy-leaching process is also applied to larger particle catalysts for continuous packed reactors. The

Table 2.5: Electrolessly plated Ni/Fe versus alloyed Ni–Fe catalysts: compositions and activity.

Catalyst variations				Sorbitol activity test	
Type	Details	Total %Ni	Surface Ni/Fe	Time (min)	Productivity
Raney® 3111	Standard	94	NA	75	1.00
Plated Ni/Fe	Plated 70°C	45	0.6	80	0.93
Plated Ni/Fe	Plated 85°C	45	0.91	75	1.00
Plated Ni/Fe	Less plated Ni	39	0.53	80	0.93
Plated Ni/Fe	Less alloy Ni	35	Not measured	90	0.83
Alloy Fe–Ni	18% Fe alloy	52	0.22	105	0.71
Alloy Fe–Ni	21% Fe alloy	42	0.12	120	0.62
Alloy Fe–Ni	27% Fe alloy	35	0.16	140	0.54
Alloy Fe–Ni	28% Fe alloy	24	0.01–0.03	110	0.68
Alloy Fe–Ni	33% Fe alloy	13	Not measured	$\gg$ 120	$\ll$ 0.63

original and still viable method uses precursor alloy granules formed from cast slabs by crushing and sieving. The size distributions so achieved have a finite width (see Figure 2.6). Pressure drop limitations led to a minimum size of 2–3 mm (effective diameter) in practice. These alloy particles are then leached into active catalysts, but with much milder conditions than for finer slurry catalyst particles, given that the goal is a much lower degree of aluminum removal. This forms a two-zone structure. The outermost zone is leached and consists primarily of Ni; unleached alloy remains in the center. Risk of attrition, i.e., breakage into fines, increases as aluminum removal increases and the porous outer zone grows in depth.

The catalysts used commercially in fixed beds are in some cases pre-activated and delivered in water, by catalyst manufacturers such as W.R. Grace. In other cases, the catalyst is made by catalyst user. Alloy granules are loaded into the hydrogenation reactor and leached *in situ*, then used immediately. A key commercial application of *in situ*-activated catalysts is butanediol synthesis by the Reppe process described previously.

Alternative alloy-forming methods have also been applied to the fixed-bed Raney catalyst approach with varying degrees of success, given performance/cost constraints [42–44].

2.4.4 Raney® Cu in synthesis of propylene glycol

The activity and versatility ranking of primary metals in Raney catalysts is Ni > Co > Cu. The use of a metal other than nickel as the primary metal component in Raney® catalysts is relatively rare. For certain niche applications where nickel is inadequate in selectivity, it is replaced by cobalt or copper. One of Raney cobalt's exemplary applications, e.g., is for selective conversion of nitriles to primary amines [45]. The cobalt catalyst is often preferred over nickel whose selectivity may be inferior even with greater usage of alkali additives. A more severe form of non-selectivity from Ni catalysts is undesired bond scission (hydrogenolysis). Examples include severed carbon–halogen bonds, largely avoidable with Raney cobalt [46] or yield loss by cleaving C–C bonds.

Overcoming C–C scission yield losses is illustrated by substituting copper for Ni in making 1,2-propanediol, or "propylene glycol," from glycerol. Glycerol (a.k.a. glycerin) became cheaply available as a by-product in producing the fatty acid methyl esters used as biodiesel fuel. As a highly oxygenated molecule, glycerol typifies starting points in many proposed "bio-renewable" processes. A broad range of nonpetroleum, i.e., agricultural and forest-based feedstocks, is upgraded or "valorized" to more useful chemicals by selective deoxygenation, consuming hydrogen. Fixed-bed Raney® Cu catalyzes the glycerol-to-glycol pathway very selectively, to over 96% yield of 1,2-diol, improving on a Cu chromite baseline (see Table 2.6).

Table 2.6: Steady-state molar yields for fixed-bed copper catalysts: glycerol to 1,2-propanediol.

Catalyst	% Conv.	1,2-Diol	1,3-Diol	EthGly	EtOH	MeOH
Raney® Cu	100	94	0.6	1.6	1.3	1.0
Raney® Cr–Cu	100	87	0.9	1.1	3.9	0.9
Cu chromite	100	85	1.0	0.6	4.4	1.0

For further details on above chemisorption, PM/Ni, plating, and propylene glycol studies, see experimental section of Ref. [47].

2.5 Concluding remarks

Raney catalysts are still viable due to their versatility, high activity and selectivity, ease of use, and rapid separations in industrial hydrogenations. The quest to connect performance of these materials to their properties has been a long, mysterious, and fascinating journey.

Acknowledgments: The author is grateful to the management at W.R. Grace for support of our research in this area and to Grace lab technicians Douglas L. Smith and Linda Wandel for their tireless work. Further, the author thanks the superb collaborators at Seton Hall University, especially Setrak Tanielyan and Prof. Robert Augustine.

References

[1] Raney, M., inventor. Murray Raney, assignee. Method of preparing catalytic material. U.S. Patent 1,563,587. 1925 Dec. 1.
[2] Sabatier, P. Vth Congress on Pure and Applied Chemistry, Berlin, 1904. IV, 663.
[3] Normann, W., inventor. Leprince and Siveke, assignee. Process for the conversion of unsaturated fatty acids or their glycerides into saturated compounds. German Patent 141,029. 1902 Aug 14.
[4] Lok, CM. The 2014 Murray Raney Award Lecture: Architecture and preparation of supported nickel catalysts. Top. Catal. 2014, 57, 1318–24.
[5] Farrauto, RJ., Bartholomew, CH. Fundamentals of industrial catalytic processes. Dordrecht, Netherlands, Springer, 1997.
[6] Internet reference. https://en.wikipedia.org/wiki/Murray_Raney. Accessed 6/20/2017
[7] Seymour, RB., Montgomery, SR. Murray Raney-Pioneer catalyst producer. In: Davis, B.H., Hettinger, WP., Jr., eds. Heterogeneous catalysis: Selected American histories. American Chemical Society Symposium Series No. 222. 1983, 34, 491–503.
[8] Lok, CM., inventor. Unilever, assignee. Ni/silica catalyst and the preparation and use thereof. U.S. Patent 5,112,792. 1989 May 16.
[9] Freel, J., Pieters, WJM., Anderson, RB. The structure of Raney Nickel: I. Pore volume. J. Catal. 1969, 14, 247.
[10] Bizhanov, FB., Fasman, AB., Sokol'skii, DV., Kozhakulov, A. Hydrogenation of phenylacetylene over Raney-nickel catalysts prepared from nickel aluminides. Int. Chem. Eng. 1976, 16, 650.
[11] Fouilloux, P. The nature of Raney nickel, its adsorbed hydrogen and its catalytic activity for hydrogenation reactions (review). Appl. Catal. 1983, 8, 1–42.
[12] Bakker, ML., Young, DJ., Wainwright, MS. Selective leaching of $NiAl_3$ and Ni_2Al_3 intermetallics to form Raney nickels. J. Mat. Sci. 1988, 23, 3921–16.
[13] Freel, J., Pieters, WJM., Anderson, RB. The structure of Raney Nickel: II. Electron microprobe studies. J. Catal. 1970, 16, 281.
[14] Fleche, G., Gallezot, P., Salome, JP. Activity and stability of promoted Raney-Nickel catalysts in glucose hydrogenation. Stud. Surf. Sci. Catal. 1991, 59, 231–6.
[15] Bartalini, G., Giuggioli, M., inventors. Montedison Fibre, assignee. Process for manufacture of hexamethylenediamine. U.S. patent 3,821,305. 1970, Oct. 22.
[16] Becker, H.-J., Schmidt, W., inventors. Bayer Aktiengesellshaft, assignee. Reduction of aromatic dinitro compounds with Raney nickel catalyst. U.S. Patent 4,287,365. 1979 June 11.
[17] Cimerol, JJ., Clarke, WM., Denton, WI., inventors. Olin Mathieson, assignee. Process of preparing aromatic polyamines by catalytic hydrogenation of aromatic polynitro compounds. U.S. patent 3,356,728. 1964, Mar. 12.
[18] Low, FG., inventor. DuPont, assignee. Butanediol preparation. Great Britain Patent 1,242,358. 1970 Feb 2.
[19] Taylor, PD., inventor. Celanese Corp., assignee. Production of tetrahydrofuran. U.S. Patent 4,064,145. 1975 Oct. 20.
[20] Dubner, WS., Shum, W., inventors. Lyondell Chemical, assignee. Production of 1,4 butanediol and/or 2-methyl 1,3 propanediol. U.S. Patent 6,969,780. 2005, Sept. 26.
[21] Huxley, EE., inventor. Phillips Petroleum Co., assignee. Sulfolene hydrogenation. U.S. Patent 6,969,780. 1980 July 14.
[22] Franczyk, TS., inventor. Monsanto Co., assignee. Process to prepare amino acid salts. U.S. patent 5,292,936. 1993 Apr. 12.
[23] Moffatt, WG. Handbook of binary phase diagrams. Schenectady, New York, USA. Genium Publishing Corp. 1986.

[24] Kordulis, C., Doumain, B., Damon, JP., Masson, J., Dallons, JL., Delannay, F. Characterization of chromium Doped Raney Nickel. Bulletin des Societe Chimiques Belges. 1985, 94(6),371–7.
[25] Macnab, GS., Anderson, RB. Structure of Raney nickel: VII. Ferromagnetic properties. J. Catal. 1973, 29, 328.
[26] Orchard, JP, Tomsett, AD., Wainwright, MS., Young, DJ. Preparation and properties of Raney nickel-cobalt catalysts. J. Catal. 1983, 84, 189.
[27] Wainwright, MS., Anderson, RB. Raney Nickel-Copper catalysts: II. Surface and pore structures. J. Catal. 1980, 64, 124.
[28] Schmidt, SR. Surfaces of Raney catalysts. In: Scaros, MJ., Prunier, ML., eds. Catalysis of organic reactions. New York, USA, Marcel Dekker, 1995, 45–59.
[29] Freel, J., Robertson, SD., Anderson, RB. Structure of Raney nickel: III. The chemisorption of hydrogen and carbon monoxide. J. Catal. 1970, 18(3),243–8.
[30] Bartholomew, CH. Hydrogen adsorption on supported cobalt, iron and nickel. Catal. Lett. 1990, 7 (1–4), 27–51.
[31] Kelly, RD. Coadsorption and reaction of H_2 and CO on Raney Nickel: Neutron vibrational spectroscopy. In: 3S'83 Symposium on Surface Science. 1983, 237–42.
[32] Montgomery, SR. Functional group activity of promoted Raney® nickel catalysts. In: Moser, W.R., ed. Catalysis of organic reactions. New York, USA. Marcel Dekker, 1984, 383–409.
[33] Rylander, PN. Hydrogenation methods. Orlando, FL, USA, Academic Press. 1985.
[34] Lieber, E., Smith, GBL. Reduction of nitroguanidine. VI: Promoter action of platonic chloride. J.A.C.S. 1936, 58(8),1417–9.
[35] Reasenberg, JR., Lieber, E., Smith, GBL. Promoter effect of platinic chloride on Raney nickel. II: effect of alkali on various groups. J.A.C.S. 1939, 61, 384–7.
[36] Schmidt, SR., inventor. W.R. Grace & Co., assignee. Promoted porous catalyst. U.S. patent 6,309,758. 1999, May 6.
[37] Harper, MJ., inventor. DuPont, assignee. Raney cobalt catalyst and a process for hydrogenating organic compounds using said catalyst. US Patent 6,156,694. 1998, Nov. 5.
[38] Mallory, GO., Hajdu, JB., eds. Electroless plating fundamentals and applications. Orlando, FL, USA. American Electroplaters and Surface Finishers Society. 1990.
[39] Morgenstern, DA. et al., inventors. Monsanto Co., assignee. Process for dehydrogenating primary alcohols to make carboxylic acid salts. US Patent 6,376,708. 2000, Apr. 11.
[40] Schmidt, SR. inventor. W.R. Grace & Co., assignee. Nickel and cobalt plated sponge catalysts. U.S. Patent 7,569,513. 2008, Apr. 7.
[41] Evans, BJ., Swartzendruber, LJ. Role of precursor alloy phases and intermediate oxides in the preparation of Raney and Urushibara iron. Hyperfine Interact. 1990, 57(1–4), 1815–22.
[42] Cheng, W-C., Lundsager, CB, Spotnitz, RM., inventors. W.R. Grace and Co., assignee. Shaped catalyst and process for making it. U.S. Patent 4,826,799 A. 1988, Apr. 18.
[43] Cheavens, T., Diffenbach, R., inventors. W.R. Grace and Co., assignee. Method for producing aluminum alloy shaped particles and active Raney catalysts from them. U.S. Patent 3,719,732 A. 1970, Dec. 17.
[44] Ostgard, D., Moebus, K., Berweiler, M., Bender, B., Stein, G., inventors. Degussa-Huels Akt., assignee. Fixed bed catalysts. U.S. patent 6,284,703. 1999, Aug. 5.
[45] Vandenbooren, F., Bosman, H., Peters, A., van den Boer, M. DSM, assignee. Process for catalytic hydrogenation of a nitrile. U.S. Patent 7,291,754. 2003, Dec 31.
[46] Lentz, C., Mullins, E., Gibson, C., inventors. Hydrogenation of halonitroaromatic compounds. Eastman Kodak Co., assignee. U.S. patent 4,929,737. 1988, Feb. 3.
[47] Schmidt, SR. Improving Raney® Catalysts through surface chemistry. In: Catalysis of organic reactions. Allgeier, AM., ed. Topics in catalysis. New York, USA. Springer. 2010, 53 (15–18), 977–1288.

Swetlana Schauermann

3 Model studies on hydrogenation reactions

3.1 Introduction

The synthesis of a large number of chemical compounds, including, e.g., pharmaceuticals and commodity chemicals, involves heterogeneously catalyzed hydrogenation of unsaturated bonds as a reaction step [1, 2]. This catalytic process has long been of persistent industrial and academic interest. Particularly over the transition metal catalysts, this type of reactions was extensively investigated in early years using both conventional catalytic methods [1, 3] and more recently surface science methodologies [2, 4–7].

Despite numerous efforts to understand the mechanisms and kinetics of hydrogenation reactions occurring over heterogeneous catalysts, there is still only limited fundamental understanding of the mechanistic details of the underlying surface processes. This lack of the atomistic insights arises mainly from the poor transferability of the results from the model studies on well-defined surfaces to the realistic hydrogenation processes. Heterogeneous catalysts employed in industrial applications are highly complex multicomponent and high surface area materials that are optimized to work for millions of turnovers, at high reaction rates and with high selectivity. They are usually composed of one or more finely dispersed catalytically active components (often on a nanometer scale), which may be transition metals or oxides, supported on a thermally stable porous oxide [1]. Additives are introduced to promote high activity, selectivity, and stability. The compositional and structural complexity of these catalysts is their principal advantage as it allows their chemical and adsorption properties to be tuned in order to maximize the activity and selectivity of a specific reaction. However, their vast chemical and structural complexity strongly hinders the mechanistic understanding of the underlying surface processes and prevents rational design of new catalytic materials. The modern surface science methodology that had emerged in the last 30 years holds a great potential to provide detailed atomistic information on the surface structure and its modification under the reaction conditions as well as to correlate these parameters to the catalytic activity. However, it cannot be applied in most cases to such complex materials as commercial heterogeneous catalysts and, as a consequence, the fundamentals of heterogeneous catalysis remain largely unexplored.

As a strategy to overcome this problem, a model catalyst approach has been developed over the last two decades by several groups [8–11]. This approach is based on the use of planar thin oxide layers, which can be epitaxially grown on single crystals. These oxide films can be used as catalysts themselves or serve as supports for metallic nanoparticles, which are deposited by physical vapor deposition under

https://doi.org/10.1515/9783110545210-003

highly controlled ultrahigh vacuum (UHV) conditions. Such nanostructured model systems allow a variety of complex features inherent to realistic catalytic surfaces to be introduced in a well-controlled fashion, while keeping them accessible for most of surface science techniques available nowadays. For this class of materials, the thermal and electrical conductivity of thin oxide films ensures an unrestricted use of surface analytical tools for detailed characterization of structural and electronic properties of model supported systems. Several examples of model catalysts have been characterized in the past by our group with respect to their geometric and electronic structure as well as their reactivity behavior [6, 7, 12–14].

There is a broad range of important structural properties of hydrogenation catalysts that need to be addressed in the atomistic-level studies. Thus, it has been recognized early on in catalysis research that the reduced dimensionality of metal particles, the presence of the low-coordinated surface sites, such as edges and corners of metallic nanoparticles, as well as the presence of the oxide support may have decisively influence the catalytic properties of heterogeneous catalysts [1]. Additionally, accumulation of carbonaceous deposits resulting from early decomposition of hydrocarbon reactants can considerably affect the activity and especially the selectivity for hydrocarbon conversions with hydrogen on transition metals [15]. These recent findings and hypotheses clearly demonstrate the need for more precise model studies over more realistic nanostructured catalysts for hydrocarbon conversions with hydrogen. In particular, studies on small metal nanoparticles supported on model planar oxide substrates under well-controlled conditions were envisaged to provide fundamental insights into the mechanisms of olefin hydrogenation at the atomic scale and elucidate the role of surface modifiers [4, 16, 17].

In this chapter, we summarize key results from our studies on the mechanisms and kinetics of olefin conversions with hydrogen over well-defined nanostructured model supported catalysts [6, 14, 18–29]. Our approach involves application of molecular beam techniques and infrared reflection–absorption spectroscopy (IRAS) to obtain detailed insights into the atomistic details of olefin hydrogenation and isomerization. As model catalysts, we employ Pd nanoparticles supported on a well-ordered Fe_3O_4 thin film, which was epitaxially grown on a Pt(1 1 1) single crystal [30]. The same types of reactions were also investigated on extended Pd single crystal surfaces to identify, which reactivity features can be traced back to the structural properties specific to Pd nanoparticles. Complementarily, nuclear reaction analysis (NRA) experiments for hydrogen depth profiling [14, 31] (in collaboration with M. Wilde and K. Fukutani, University of Tokyo) and theoretical calculations [29, 32, 33] provided important information on the details of hydrogen interaction with Pd.

Two classes of the surface hydrogenation reactions will be discussed in this chapter: hydrogenation of simple olefins and partial selective hydrogenation of α, β-unsaturated ketones and aldehydes to unsaturated alcohols.

3.2 Hydrogenation of simple olefins

3.2.1 *cis*-2-Butene hydrogenation and *cis–trans* isomerization over model supported Pd nanoparticles

According to the generally accepted Horiuti–Polanyi reaction mechanism [34], *cis*-2-butene conversions with hydrogen over transition metal catalysts proceed via a series of consecutive hydrogenation and dehydrogenation steps:

cis-2-butene $\xrightarrow{+D}$ Butyl → (−H) *Trans*-2-butene-d_1; (+D) Butane-d_2

If D_2 is used as a hydrogenation reactant, a 2-butyl-d_1 intermediate is formed in the first half-hydrogenation step, which is a common reaction intermediate (RI) for the *cis–trans* isomerization, the H/D exchange, and the hydrogenation reaction pathways. The 2-butyl-d_1 intermediate can undergo β-hydride elimination resulting in alkene formation, producing either the original molecule or a *cis–trans*-isomerized molecule *trans*-2-butene-d_1 [35]. With D_2 being one of the reactants, each *cis–trans* isomerization event is accompanied by substitution of one hydrogen atom with a deuterium (H/D exchange), allowing for a distinction between the reactant *cis*-2-butene and the product *trans*-2-butene-d_1 in the gas phase by using mass spectrometry. Alternatively, a second hydrogen (deuterium) atom can be inserted into the carbon–metal bond of the 2-butyl-d_1 intermediate leading to formation of butane-d_2. The adsorbed alkene as well as the 2-butyl-d_1 species may also dehydrogenate yielding a variety of carbonaceous surface species, such as alkylidynes and or strongly dehydrogenated carbonaceous species stoichiometrically close to carbon [2, 4].

There is an ongoing discussion about possible involvement of subsurface hydrogen species (absorbed below the outermost surface layer or in the volume of metal nanoparticles) in olefin hydrogenation. The traditional opinion that only surface hydrogen species participate in hydrogenation process [1] was questioned for the first time by Ceyer and coworkers studying ethylene hydrogenation on Ni(1 1 1) [36]. More recent studies on supported nanoparticles provided the first experimental evidence that the weakly bound subsurface/volume-absorbed hydrogen species can be crucial for hydrogenation [16]. Specifically, a high hydrogenation activity was observed under low-pressure conditions on supported Pd clusters but not on the singles crystal, which was attributed to the unique ability of the small particles to store large amounts of hydrogen atoms in a confined subsurface volume near the metal–gas interface. This scenario can potentially result in some important implications for the overall activity and selectivity of the hydrogenation catalyst: if one of the key reaction participants – subsurface hydrogen – can be only slowly replenished under reaction conditions, the hydrogen permeability of the metal surface can play a crucial role in the reaction kinetics. The potential involvement of subsurface hydrogen in hydrogenation of the double bond remained a controversial issue for a long time [37], mainly because of the lack of experimental techniques capable of separately monitoring different hydrogen species under the reaction conditions.

In our recent studies on well-defined supported model catalysts, we were able to solve these problems and unravel the microscopic details of olefin hydrogenation. Here, we will briefly summarize the most important results on hydrogenation and *cis–trans* isomerization of *cis*-2-butene obtained in these experiments [14, 18–23, 29, 32, 33].

Figure 3.1 shows the scanning tunneling microscopy image (a) and the model of Pd nanoparticles employed in this study. The nanoparticles were supported on thin planar Fe_3O_4 film, which was prepared *in situ* under UHV conditions on a planar Pt(1 1 1) substrate [38]. The Pd nanoclusters are on average 6 nm in diameter and exhibit mainly (1 1 1) facets (~80%) and a smaller fraction of (1 0 0) facets and other low-coordinated surface sites, such as edges and corners (~20%). It has been spectroscopically shown by using carbon monoxide as a probe molecule for different adsorption sites (Figure 3.1a) that when a sub-monolayer amount of C is produced on the Pd nanoclusters, it selectively blocks the low-coordinated surface sites, such as edges, corners, and (1 0 0) facets of the clusters, while leaving the majority of the regular (1 1 1) facets C-free [39]. We used both types of catalyst – C-free and C-modified – to compare their catalytic activities and selectivities with regards to the competing *cis–trans* isomerization and hydrogenation reaction routes. Carbon was deposited onto Pd nanoparticles by thermal decomposition of *cis*-2-butene [18] and the spatial distribution of the resulting carbonaceous deposits was characterized by IRAS using carbon monoxide as a probe molecule (Figure 3.1a).

The formation rates of the two primary reaction products – *trans*-2-butene-d_1 and butane-d_2 – were probed under isothermal conditions using molecular beam techniques in the temperature range between 190 and 260 K. In this series of experiments, the sample was continuously exposed to a D_2 beam at a high flux (for details, see Ref. [18]), while *cis*-2-butene was pulsed at a much lower flux (flux ratio of D_2/*cis*-2-butene = 760) starting 90 s after the beginning of the D_2 dosing. The delay in the olefin dosing was made in order to ensure that both the surface and the volume of Pd nanoparticles were saturated with deuterium. A typical MB sequence included 50 short butene pulses (4 s on, 4 s off time) to probe the transient reactivity followed by 30 longer pulses (20 s on, 10 s off time) to probe the steady-state behavior. The time evolution of the reaction rates measured at 260 K for formation of *trans*-2-butene-d_1 (black curves) and butane-d_2 (gray curves) is displayed in Figure 3.2 for the pristine (Figure 3.2a) and C-modified (Figure 3.2b) Pd nanoparticles. The start of the *cis*-2-butene exposure corresponds to zero on the time axis.

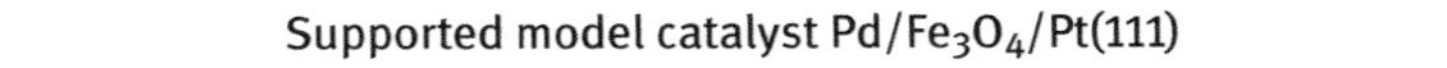

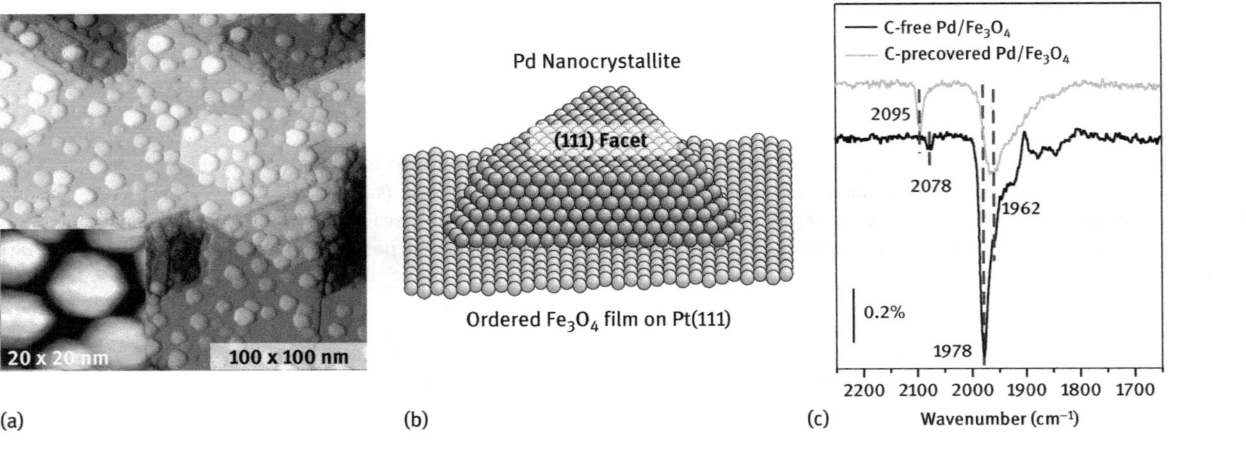

Figure 3.1: (a) Scanning tunneling microscopy (STM) image of the Pd nanoparticles supported on Fe_3O_4/Pt(1 1 1) used in the experiments described here, (b) together with a schematic representation of their structure (from Ref. [38]). (c) IR spectra of CO adsorbed on (i) pristine and (ii) C-precovered supported Pd nanoparticles. The data indicate that carbon is non-uniformly distributed over the Pd particles and blocks preferentially low-coordinated sites, such as edges, corners, and bridge sites on (1 0 0) facets. The majority of the surface sites on the regular (1 1 1) terraces remain unaffected.

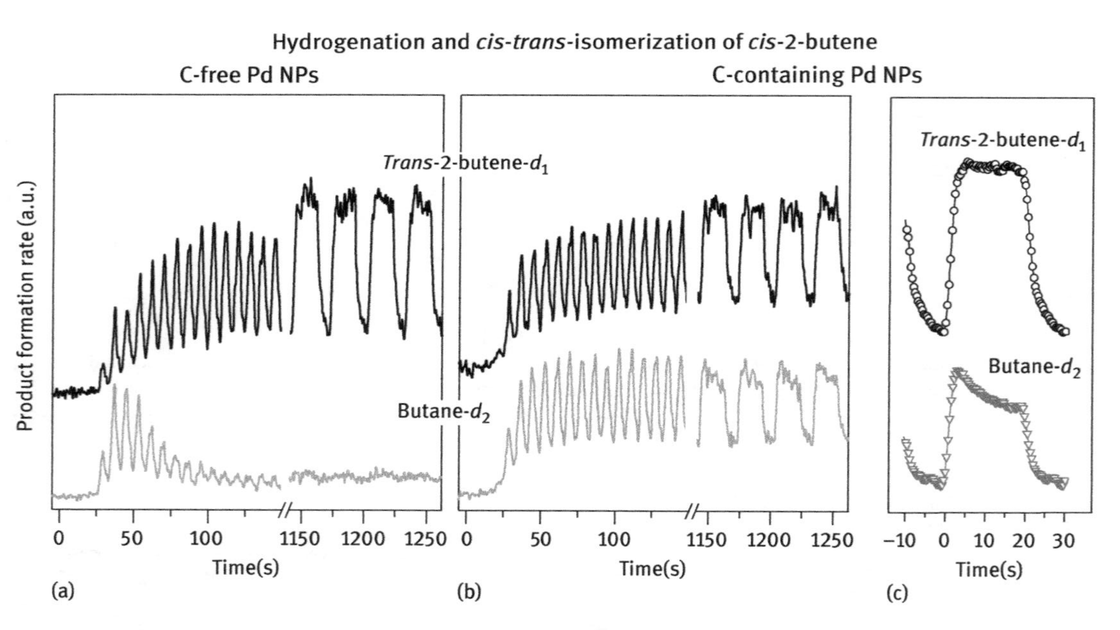

Figure 3.2: Results from isothermal pulsed molecular beam experiments on the conversion of *cis*-2-butene with D_2 at 260 K on (a) initially clean and (b) C-modified $Pd/Fe_3O_4/Pt(1\,1\,1)$ model catalysts (after Ref. [20]). Shown is the evolution of the reaction rates as a function of time for *trans*-2-butene-d_1 (black curves) and butane-d_2 (gray curves). The catalysts were exposed to D_2 beam continuously, while the *cis*-2-butene beam was pulsed; (c) average steady-state reaction rates calculated for *cis–trans* isomerization (black curve) and hydrogenation (gray curve) from the last 30 long pulses of the dataset show in (b).

On the pristine and C-modified Pd nanoparticles (Figure 3.2a), both reaction pathways exhibit a short induction period followed by a transient period of high activity [18]. However, on the pristine Pd clusters, only *cis–trans* isomerization activity is maintained under the steady-state conditions, while the hydrogenation rate quickly decreases to zero. On contrast, not only *cis–trans* isomerization but also hydrogenation is sustained in steady state on the catalyst that was modified with carbon prior to reaction (Figure 3.2b). This unique catalytic behavior clearly demonstrates that the persistent hydrogenation activity can be maintained on Pd nanoparticles only if carbon is present on the surface at the low coordinated sites.

Detailed kinetic analysis of both reaction rates strongly suggests that two nonequivalent D species must be involved into the first and the second half-hydrogenation steps to produce butane-d_2 [14, 18, 19, 21, 22]. Briefly, the sustained *cis–trans* isomerization rate evidences the fast formation of the common butyl-d_1 RI and with this shows that sufficient amounts of both the adsorbed *cis*-2-butene species and the surface D atoms are available under the reaction conditions. However, the availability of these D atoms, which are involved into the first half-hydrogenation step, is apparently not a sufficient condition for the second half-hydrogenation to form butane-d_2. It can be concluded from this observation that the presence of a different special type of D atom (further indicated as D*) must be required for the second half-hydrogenation step. These D*-species are obviously present at the beginning of reaction on the both types of Pd particles pre-saturated with deuterium, which gives rise to initially high hydrogenation rates. However, they cannot be replenished under steady-state conditions on pristine Pd particles, which lead to selective suppression of hydrogenation.

A similar conclusion on the two types of D species involved in the reaction can be drawn from the analysis of the pulse profiles obtained under steady-state conditions for both products formed over C-modified nanoparticles (Figure 3.2c) [19, 23]. Here, the averaged reaction rates are shown, obtained over C-modified Pd particles upon modulation of the olefin beam. Prior to each pulse, the surface was exposed to a continuous D_2 beam in order to completely re-saturate the particles with D species so that the reactivity immediately after resuming the *cis*-2-butene beam corresponds to the reactivity of the catalyst saturated with deuterium. Two different pulse profiles were observed for *cis–trans* isomerization (upper trace) and hydrogenation (lower trace). While the isomerization rate quickly reaches its steady-state value and then remains constant during the duration of pulse, the hydrogenation rate shows the highest reactivity at the beginning of the pulse but then decreases to a significantly lower value upon continuing olefin exposure. This behavior can be explained only by assuming that two different types of D species are involved into the first and second half-hydrogenation steps. Indeed, the constant rate of *cis–trans* isomerization points to constant surface concentrations of the reactants – both the RI (butyl-d_1) and the surface D. In contrast, the decreasing hydrogenation rate unambiguously means that the surface concentration of one of the RI must decrease during the duration of the pulse. As the competing reaction pathways differ only in the second half-hydrogenation step, this reactant can be only the D* species, which is involved into the second half-hydrogenation. In agreement with the discussion above, the clearly different pulse profiles observed can be explained only by assuming that nonequivalent D species participate in the first and second half-hydrogenation steps of *cis*-2-butene.

In the following, we will address two closely related questions: what is the nature of this special type of D* species, and how does carbon promote the sustained hydrogenation activity?

There is general agreement in the literature that two different types of H(D) species can be formed on Pd:H(D) species adsorbed on the surface that are formed in a nonactivated dissociative adsorption and the subsurface/volume-absorbed H(D) species (in the following, we will denote the subsurface/volume-absorbed state as subsurface state for simplicity), which have to overcome an activation barrier to diffusion into the subsurface region of Pd [14, 31, 40]. The spatial distribution of H atoms on Pd – both single crystals and supported Pd nanoparticles – was recently investigated by hydrogen depth profiling via $^{1}H(15N, \alpha\gamma)^{12}C$ NRA [14, 31]. In addition, this method allows independent quantification of the abundance of surface and subsurface H species as a function of gas-phase hydrogen pressure. The most important observation of these studies was that the concentration of the subsurface H species shows a strong H_2 pressure dependence up to at least 2×10^{-5} mbar, whereas the coverage of surface H species saturates as much lower pressures (below 1×10^{-6} mbar) does not change upon further H_2 pressure increase. These findings are in agreement with the results of other previous experimental and theoretical studies which indicate that H binds much stronger on the surface (~0.8 eV per Pd atom) as compared to the subsurface (~0.4 eV) [41]. These results indicate that the strongly binding surface adsorption state populates first and the population of the subsurface hydrogen depends more strongly on the H_2 pressure in the environment. Additionally, even small variations of H_2 gas pressure were measured to result in considerable changes of subsurface H concentration, which are comparable to the total number of surface-adsorbed H atoms.

In our experimental approach, these experimental observations made by NRA for hydrogen depth profiling [31] were complemented by the reactivity measurements in relaxation-kinetics molecular beam experiments [14], where the reaction rates of the competing *cis*–*trans* isomerization and hydrogenation pathways were investigated as a function of the D_2 pressure.

The results of these experiments on the C-modified Pd nanoparticles are displayed in Figure 3.3. Prior the experiment, Pd nanoparticles were preexposed to D_2 and *cis*-2-butene to reach the steady-state regime, and then the D_2 beam was interrupted for 100 s until all D species – both surface and subsurface – were removed either via desorption or reaction, as indicated by vanishing reaction rates. Thereafter, the D_2 beam was resumed and the evolution of the hydrogenation (butane-d_2) and the isomerization (*trans*-2-butene-d_1) products was monitored as a

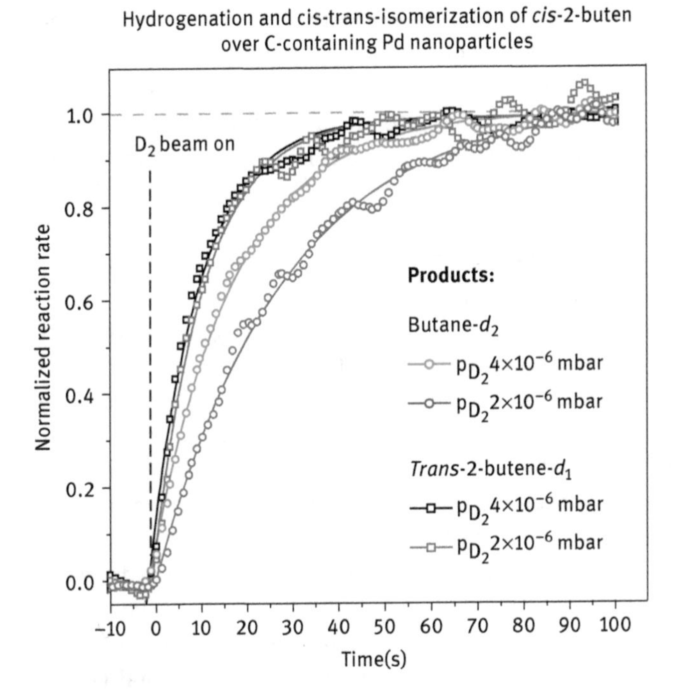

Figure 3.3: Time evolution of the normalized reaction rates for the *cis–trans* isomerization (*trans*-2-butene-d_1) and hydrogenation (butane-d_2) products in the *cis*-2-butene + D_2 reaction after a temporary intermission of the D_2 beam. The reaction rates are obtained at 260 K over the C-modified $Pd/Fe_3O_4/Pt(1\,1\,1)$ catalyst at different D_2 pressures – 2×10^{-6} and 4×10^{-6} mbar – with the ratio *cis*-2-butene:D_2 kept constant (after Ref. [14]).

function of time. The main idea of this experiment was to repopulate different D species – surface and subsurface ones – with different time constants, which will strongly depend on the D_2 pressure in the gas phase, and to monitor how fast the reaction rates of both competing reaction pathways return to their steady-state value. The time evolution of both reaction products in this case will be limited by the formation rate of various D species. In view of the different binding energies of surface and subsurface H(D) species and existing activation barrier for H(D) diffusion into the subsurface region, it can be expected that steady-state concentration of surface D species will be reached first, while the concentration of subsurface D will return to its steady-state value with some delay. In addition, the strong dependence of the subsurface H(D) concentration on the $H_2(D_2)$ pressure suggests that the characteristic time for replenishing the subsurface reservoir should be strongly dependent on the $D_2(H_2)$ pressure.

The relaxation-kinetics molecular beam experiments were conducted for two different D_2 pressures (4×10^{-6} and 2×10^{-6} mbar) with a constant *cis*-2-butene:D_2 ratio of

2×10^{-3} [14]. All reaction rates at both pressure conditions returned to exactly the same steady-state values after the D_2 beam was resumed. However, the transient time evolution of different reaction rates (*cis–trans* isomerization vs. hydrogenation) exhibits very different behavior as well as different dependence on the D_2 pressure. At both D_2 pressures, the isomerization rate returns to the steady-state level with short and nearly identical time constants: $\tau_{cis-trans} = 11.0 \pm 0.3$ and 11.4 ± 0.3 s. In contrast, the evolution of the hydrogenation rate during relaxation is considerably slower and exhibits a strong pressure dependence with characteristic times of $\tau_{\mathrm{hydr}} = 18.3 \pm 0.3$ and 28.3 ± 0.5 s for 4×10^{-6} and 2×10^{-6} mbar of D_2, respectively.

The fast and pressure-independent characteristic times of the isomerization rates $\tau_{cis-trans}$ suggest that this reaction route is linked to the abundance of surface D atoms. In contrast, the prominent D_2 pressure dependence of the hydrogenation rate suggests that the second half-hydrogenation of butyl-d_1 to butane-d_2 is strongly sensitive to the D_2 pressure. In combination with the NRA results, this observation leads us to the conclusion that subsurface D(H) species are required for the second half-hydrogenation step. Note that the D species required for hydrogenation cannot be the surface-adsorbed D, as in that case the characteristic time of the hydrogenation product τ_{hydr} would be pressure independent in the same way as the characteristic time of the isomerization pathways $\tau_{cis-trans}$. The significantly slower recovery of the hydrogenation rate as compared to the isomerization rate is a natural consequence of the fact that the surface D atoms have to diffuse into subsurface of Pd nanoclusters to form subsurface D species, which is a slow and activated process, whereas D_2 dissociation on the Pd surface is not [40]. With this, it can be concluded that the D* species discussed above as a special sort of D required for full hydrogenation can be related to subsurface D.

It should be noted that involvement of subsurface hydrogen into the second half-hydrogenation step does not necessarily imply that this species is directly involved in attacking the metal–carbon bond. Instead, the electronic and/or adsorption properties of surface D(H) species could be modified in presence of subsurface D(H) so that it becomes more prone for attacking the metal–carbon bond. Recent theoretical study on hydrogenation of alkyl compounds on Pd by surface versus subsurface hydrogen corroborates this hypothesis [32]. In this study, subsurface H was shown to strongly lower the binding energy of the surface-adsorbed H species, which results in a strong decrease of the activation energy for insertion of surface H into the Pd–C bond of alkyl species and, as a consequence, strong facilitation of hydrogenation rate. In absence of subsurface H species, the binding energy of surface H is significantly higher and the corresponding activation energy of alkyl hydrogenation by surface H is considerably greater.

Based on these findings, we can now explain the observations displayed in Figure 3.2. On the pristine Pd nanoparticles pre-saturated with D_2 prior to olefin exposure, both surface and subsurface D species are populated, which results in high initial rates of both *cis–trans* isomerization and hydrogenation. After the prolonged *cis*-2-butene exposure, the hydrogenation route becomes selectively suppressed because the subsurface D species are depleted and cannot be replenished fast enough on the time scale of the experiment to ensure sustained hydrogenation activity. The inability to quickly repopulate subsurface D arises most likely from hindered D subsurface diffusion on the surface covered with hydrocarbons. On the other hand, sustained hydrogenation observed of the C-modified Pd nanoparticles suggests that the replenishment of subsurface D species occurs at a much faster rate on the C-containing Pd nanoparticles. There is only one conceivable way to rationalize this observation: the carbonaceous species adsorbed at the low-coordinates sites of Pd clusters (edges and corners) must strongly facilitate D diffusion from the surface into the particle volume. The latter hypothesis was corroborated both by the theoretical calculation [29, 32, 33] and experiment [21, 42], which will be shortly summarized below.

3.2.2 Subsurface hydrogen diffusion on pristine versus C-modified Pd nanoparticles: Theory and experiment

Recent theoretical calculation provided detailed atomistic insights into carbon-induced facilitation of hydrogen subsurface diffusion. By calculating the activation barriers for the H subsurface diffusion process, it was possible to show carbon deposition results in a pronounced expansion of the Pd lattice near the particle edges [29, 33]. This structure modification nearly lifts the activation barrier for hydrogen subsurface diffusion through these newly formed broad "channels" and with this strongly enhanced H subsurface diffusion rate. In contrast, the lateral rigidity of the extended Pd(1 1 1) surface hinders the lattice expansion and therefore carbon deposition showed no effect on the activation barrier for hydrogen subsurface diffusion on this surface. These theoretical predictions are in an excellent agreement with the experimental observations, showing that C deposition allows maintaining of the hydrogenation rate on atomically flexible Pd nanoclusters but not on the laterally rigid Pd(1 1 1) surface [22]. These computational results prove the conceptual importance of the atomic flexibility of sites near particle edges on Pd nanoparticles, which plays an exceptionally important role in H subsurface diffusion.

Finally, the rate of hydrogen subsurface diffusion was addressed experimentally by performing $H_2 + D_2 \rightarrow HD$ reaction over pristine and C-modified Pd nanoparticles [21, 42]. In order to probe the rate of subsurface diffusion, this reaction was conducted in the temperature range where it is limited by the formation rate of subsurface H(D) species: between 200 and 300 K. Generally, formation of HD can proceed either via recombination of two surface H and D species (above 300 K) or via recombination of one surface and one subsurface H or D species (between 200 and 300 K) as it was shown in previous NRA experiments combined with temperature programed desorption [31].

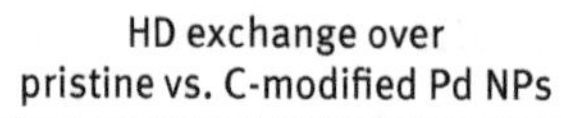

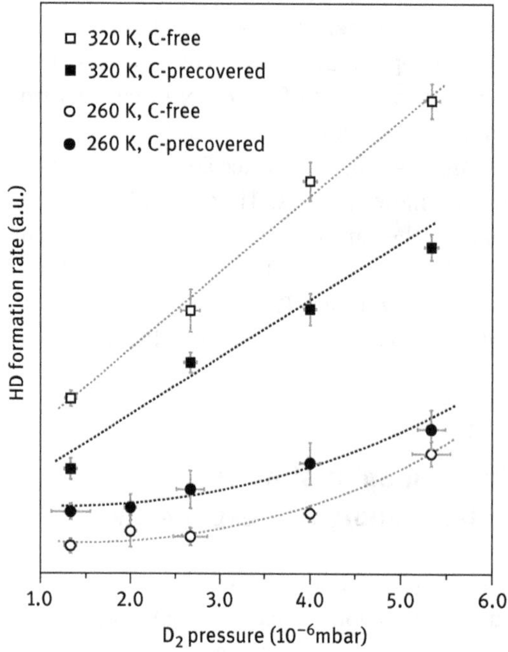

Figure 3.4: The steady-state HD formation rates obtained on the pristine and C-precovered Pd nanoparticles supported on Fe_3O_4/Pt(1 1 1) at 260 and 320 K in the D_2 pressure range from 1.3 × 10^{-6} to 5.3 × 10^{-6} mbar at the constant reactant ratio D_2:H_2 = 71. C deposition results in the ~30% decrease of the HD formation rate at 320 K and in ~100% increase of the reaction rate at 260 K (after Ref. [21]).

Figure 3.4 shows the steady-state HD formation rates measured during D_2 and H_2 exposure for the reaction temperatures 260 and 320 K on clean and C-modified Pd nanoparticles for a range of D_2 pressure conditions with a constant D_2:H_2 ratio. The plot shows the steady-state formation rates established after the initial transient period. Carbon deposition affects the HD formation rate in a dramatically different way for the two investigated reaction temperatures: while at 320 K the overall reaction rate *is reduced* by about 30% on C-containing particles [18], the reaction rate *increases* by nearly 100% at 260 K on the C-modified particles for all pressures studied. The 30% decrease of the HD formation rate at 320 K, at which HD formation is dominated by the recombination of the surface H and D species, is merely a consequence of the blocking of surface adsorption sites by deposited carbon. The magnitude of this effects is in a good agreement with the total area covered by carbon (20%) estimated from the absolute amounts of low-coordinated sites and regular (1 1 1) terraces.

Interestingly, even though 20% of the surface is blocked by C, the HD formation rate increases by factor of two on this surface at 260 K, at which desorption (e.g., H and D recombination) involves at least one subsurface H(D) species. This effect can be explained only by the higher formation rate of the subsurface H(D) species on the C-modified particles resulting in a higher steady-state concentration of H(D) in the subsurface. Note that the H(D) surface concentration under this reaction conditions equals to unity. At the microscopic level, facilitation of the subsurface H diffusion through the C-modified low-coordinated sites is most likely even more pronounced. Indeed, the nearly 100% increase of the overall subsurface diffusion rate arises from modification of only the ~20% of the surface sites constituting edges and corners of Pd particles. This means that the local diffusion rate through these C-modified sites increases by at least an order of magnitude as compared to the clean surface.

The major conclusions of this study are summarized in Figure 3.5. First, we provide the direct experimental evidence that the presence of subsurface hydrogen is required for the hydrogenation pathway. Particularly, the subsurface hydrogen is involved – directly or indirectly – into the second half-hydrogenation step of butyl to butane. The *cis–trans* isomerization pathway can be maintained in the presence of surface-adsorbed H(D) species. Second, we show that the hydrogenation pathway could be maintained over long periods of time only over Pd nanoparticles, at which the low-coordinated surface sites, such as edges and corners, were modified by deposition of strongly dehydrogenated carbonaceous deposits/carbon, but not on the pristine particles. This promoting effect of C on Pd nanoparticles can be ascribed to pronounced facilitation of subsurface hydrogen diffusion through C-modified low-coordinated surface sites, which allows for effective replenishment of the subsurface reservoir H(D) species. Computational results support this

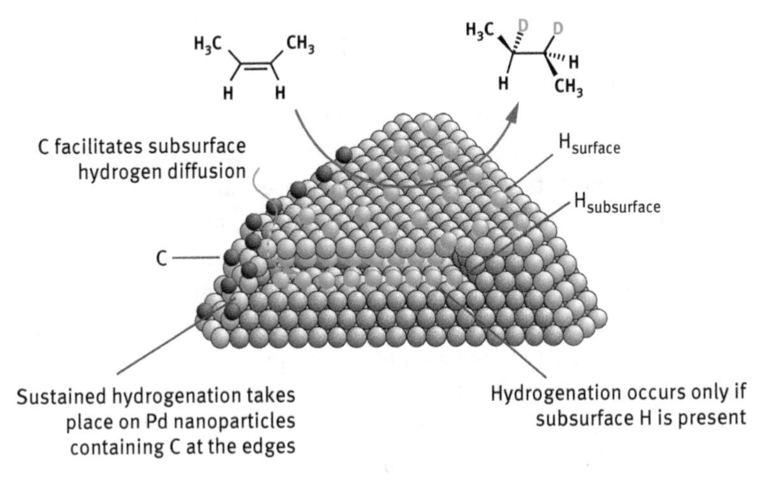

Figure 3.5: Proposed reaction mechanism of olefin hydrogenation on Pd nanoparticles.

hypothesis and show that carbon deposition near particle edges results in a lattice expansion and a concomitant elimination of the activation barrier for hydrogen subsurface diffusion.

Experimentally, the rate of the subsurface hydrogen diffusion was addressed by the $H_2 + D_2 \rightarrow HD$ exchange reaction in the low-temperature regime, at which this reaction is limited by the formation rate of subsurface H(D) species. In agreement with theoretical predictions, we obtained an experimental evidence of a strong facilitation of H subsurface diffusion rate through C-modified low-coordinated sites as compared to C-free particles. Our experimental observations point to an exceptional importance of the atomically flexibility of low-coordinated surface sites on Pd particles in the hydrogen diffusion process and identify carbon as a crucial participant in surface reactions, which governs the rate of subsurface hydrogen diffusion and with this the overall hydrogenation rate.

3.3 Partial selective hydrogenation of α,β-unsaturated ketones and aldehydes to unsaturated alcohols

Selective partial hydrogenation of α,β-unsaturated aldehydes and ketones to unsaturated alcohols is of broad interest for numerous industrial applications including synthesis of fine chemicals and pharmaceuticals [43, 44]. The undesired hydrogenation of the C=C bond to form the saturated aldehyde is, however, a thermodynamically more favored process with the saturated ketones being by about 30 $kJ.mol^{-1}$ more stable than the unsaturated alcohols [45]. Hence, selective hydrogenation of unsaturated ketones and aldehydes to unsaturated alcohols requires manipulation of the reaction kinetics by appropriate catalysts. Previously, this class of reactions was widely investigated in catalytic studies on powdered materials and model single crystal studies, including Pt, Pd, Rh, Ni, Cu, Ag, and Au as active component [43, 44, 46–52]. Particularly over Pt group metals, the undesired hydrogenation of the C=C bond in α,β-unsaturated aldehydes is strongly favored over the desired C=O bond hydrogenation with the selectivity close to 100%. Different structural parameters were put forward as key components controlling the chemoselectivity, including the amount of steric hindrance to adsorption via the C=C bond [53], alloying with other metals [54] and addition of surface modifiers [55, 56]. The selectivity to C=O bond hydrogenation can be improved by using ketones or aldehydes with sterically protected C=C groups, e.g., prenal instead of acrolein [53, 57]. Complementary, partially reducible oxide supports, such as TiO_2, were employed to provide Lewis-acid sites, which were discussed to coordinate and thus activate the C=O bond [58–60]. Additionally, a large body of theoretical work and experimental mechanistic studies on single crystal surfaces has been reported, in which the parameters controlling the chemoselectivity of this reaction were addressed [61–64].

Despite these efforts, the problem of the selective hydrogenation of a C=O bond in α,β-unsaturated ketones and aldehydes remains unresolved, particularly for the smallest molecules, such as acrolein, possessing no functional groups to protect the C=C bond. Also, the nature of the adsorbates formed on the catalytic surface under the reaction conditions, including the RIs, is not fully understood. This microscopic-level information on the underlying surface processes is crucial for achieving rational design of selective catalysts for this class of reactions.

3.3.1 Kinetics of partial acrolein hydrogenation over Pd nanoparticles versus Pd(1 1 1)

In this section, we will provide an overview of the recent mechanistic studies from our group on partial selective hydrogenation of acrolein over Pd, in which a detailed atomistic picture of this reaction was obtained. This catalytic process was investigated on two types of well-defined surfaces prepared *in situ* under UHV conditions: (i) a single crystal Pd(1 1 1) and (ii) Pd nanoparticles supported on a flat model Fe_3O_4/Pt(1 1 1) oxide support [24, 26, 28]. These surfaces were exposed to the reactants – acrolein and H_2 – in a highly controlled way by using molecular beam techniques and under isothermal reaction conditions. Simultaneously, the formation rate of the gas-phase products was monitored by quadrupole mass spectrometry (QMS) and the evolution of the surface species formed on the surface turning over was followed by *in-situ* IRAS. The following scheme shows the possible reaction pathways of acrolein partial hydrogenation.

Propenol
+ H_2
OH
OH
+ H_2
O
Acrolein
+ H_2
O
Propanal
+ H_2
OH
Propanol

Figure 3.6 shows the formation rates of two competing reaction pathways leading to selective hydrogenation of either the C=C double bond to form propanal (black traces, Figure 3.6a, b) or the C=O double bond to form propenol (blue traces, Figure 3.6c or d) measured on supported Pd nanoparticles (left side) and Pd(1 1 1) (right side) as a function of temperature. In these experiments, the surface was exposed continuously to a H_2 molecular beam prior to acrolein exposure to ensure Pd saturation with hydrogen. At time moment indicated as zero, a series of acrolein pulses was

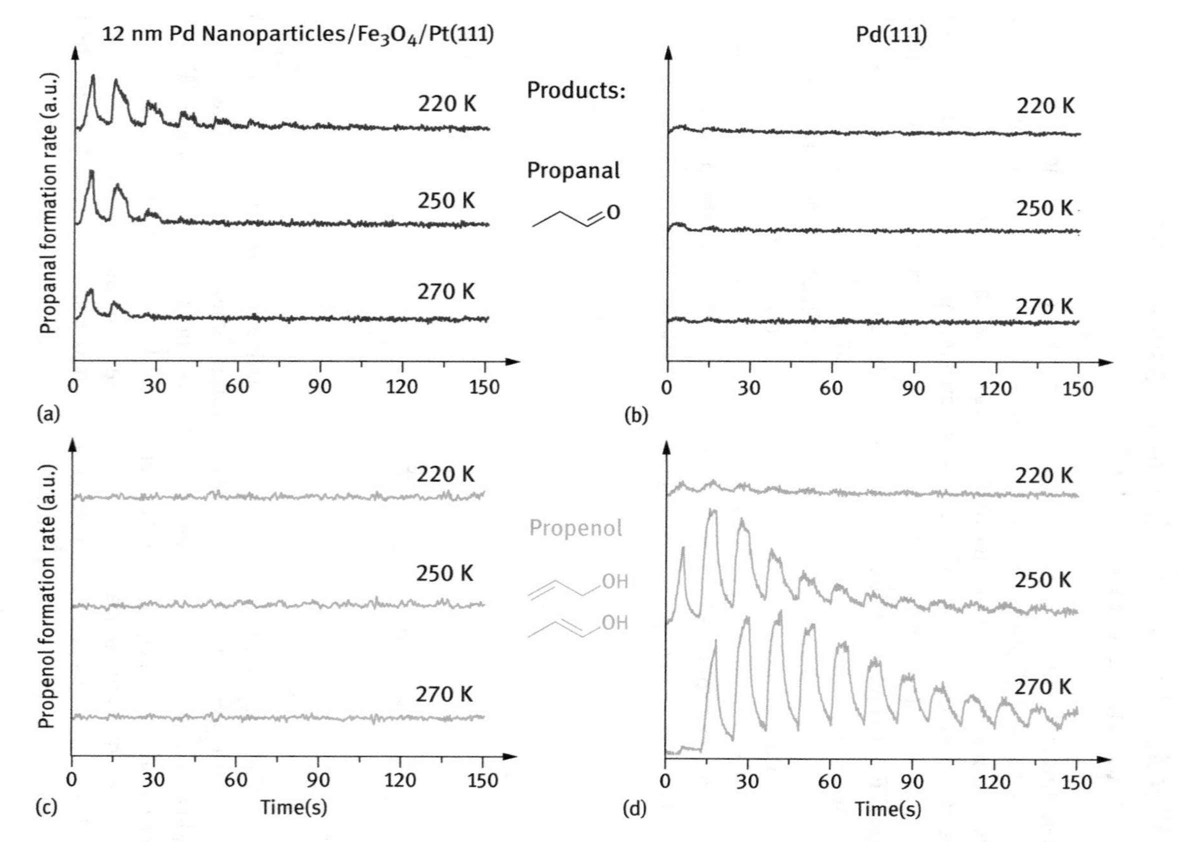

Figure 3.6: Formation rates of the reaction products – propanal (upper row) and propenol (lower row) – on 12-nm-sized supported Pd nanoparticle (a, c) and Pd(1 1 1) (b, d) during continuous dosing of H_2 and pulsed dosing of acrolein at different temperatures (after Ref. [24]).

applied and the formation rates of reaction products were recorded by mass spectrometry. On the Pd nanoparticles, only one product – (undesired) propanal – was formed in the first few pulses with the reaction rate quickly dropping to zero. No formation of propenol was observed for supported Pd nanoparticles. These observations agree with the earlier studies performed under ambient pressure conditions on powdered supported Pd catalysts, showing almost exclusive hydrogenation of the C=C bond in acrolein [44, 65]. In sharp contrast to Pd nanoparticles, very high catalytic activity toward the desired reaction product propenol was recorded on Pd (1 1 1) (Figure 3.6d). In addition, nearly no propanal was formed on this surface, which means that Pd(1 1 1) exhibits 100% selectivity toward propenol formation. The propenol formation rate exhibits a very strong temperature dependence with a maximum of conversion between 250 and 270 K. In our studies, no steady-state regime could be achieved: at all investigated temperatures, the reaction rate passes through a maximum and then decreases to zero. In order to obtain more detailed information on the composition of the catalytically active surface, we performed time-resolved IRAS measurements under the reaction conditions using a continuous exposure of both hydrogen and acrolein via independent molecular beams. The evolution of both reaction products recorded in this experiment is displayed on Figure 3.7. Here, the propanal formation rate (black trace) remained always zero. The time dependence of the propenol formation rate at 270 K suggests (blue trace)

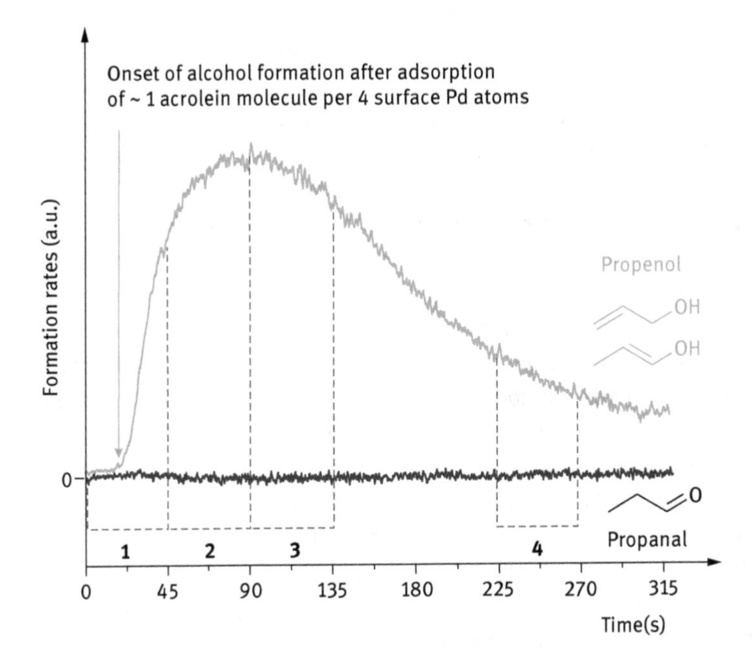

Figure 3.7: Formation rate of propenol (blue line) and propanal (black line) on Pd(1 1 1) at 270 K under continuous exposure of H_2 and acrolein (after Ref. [24]).

three reactivity regimes: (i) an induction period at the beginning of the reaction when acrolein is accumulated on the surface but no propenol is formed, (ii) the period of highest activity and selectivity toward propenol formation, and (iii) deactivation of the catalysts leading to a slow decrease of the reaction rate.

3.3.2 Identification of the adsorbed surface species on Pd(1 1 1) under the reaction conditions by IRAS

Figure 3.8 shows three representative IR spectra obtained in the three reactivity regimes on the surface turning over. The spectrum (2) was obtained on the surface during the first 45 s, comprising the induction period and the period of growing reactivity. The spectrum (3) was recorded during the next 45 s (from 45 to 90 s after the beginning of acrolein exposure), corresponding to the high propenol formation rates. The bottom spectrum (4) shows the composition of the deactivated surface at the end of the experiment (obtained between 450 and 540 s). In these spectra, three spectral regions characteristic for the CH_x stretching vibrations (3,200–2,700 cm^{-1}), C=O and C=C stretching vibrations (1,850–1,550 cm^{-1}), and CH_x deformation as well as C–O and C–C stretching vibrations (≤1,500 cm^{-1}) can be distinguished.

For comparison, the IR spectrum (1) of an acrolein monolayer adsorbed on Pd(1 1 1) at 100 K, which was obtained after exposure of 3.6×10^{14} acrolein molecules.cm^{-2}, is displayed at the top of Figure 3.8. At this temperature, acrolein does not react and adsorbs molecularly on Pd. The spectrum of molecularly adsorbed acrolein was discussed in detail previously [66, 69]. The most pronounced band appears at 1,663 cm^{-1}, which can be related to the stretching vibration of the carbonyl (C=O) group that is conjugated to the C=C group, with a less pronounced band in the 1,430–1,400 cm^{-1} range, which is assigned to a scissor deformation of the methylene (CH_2) group.

The spectrum (2) in Figure 3.8 obtained during the induction period and the period of growing reactivity is clearly different from that of molecularly adsorbed acrolein. A pronounced IR vibrational mode appears at 1,755 cm^{-1} and a second one near 1,120 cm^{-1}. The vibration at 1,755 cm^{-1}, which is blue-shifted by 92 cm^{-1} relative to the carbonyl band in acrolein, is typical for the carbonyl stretching mode in saturated aldehydes and ketones [70, 71] and is associated with a C=O stretching vibration that is not conjugated to a C=C group. The appearance of this new vibration under reaction conditions points to the formation of an oxopropyl surface species, resulting from partial hydrogenation of the C=C group in acrolein with only one H atom. One of two possible structures of this adsorbate is shown on the right of Figure 3.8. In the course of the reaction, the intensity of the vibrational band at 1,755 cm^{-1} grows and later saturates (see spectra 3 and 4), even when the reaction rate of propenol formation in the gas phase decreases to zero. This observation means that the oxopropyl species is not the direct RI for propenol formation but rather a spectator species. We will refer to it as to the spectator species S1 in the following.

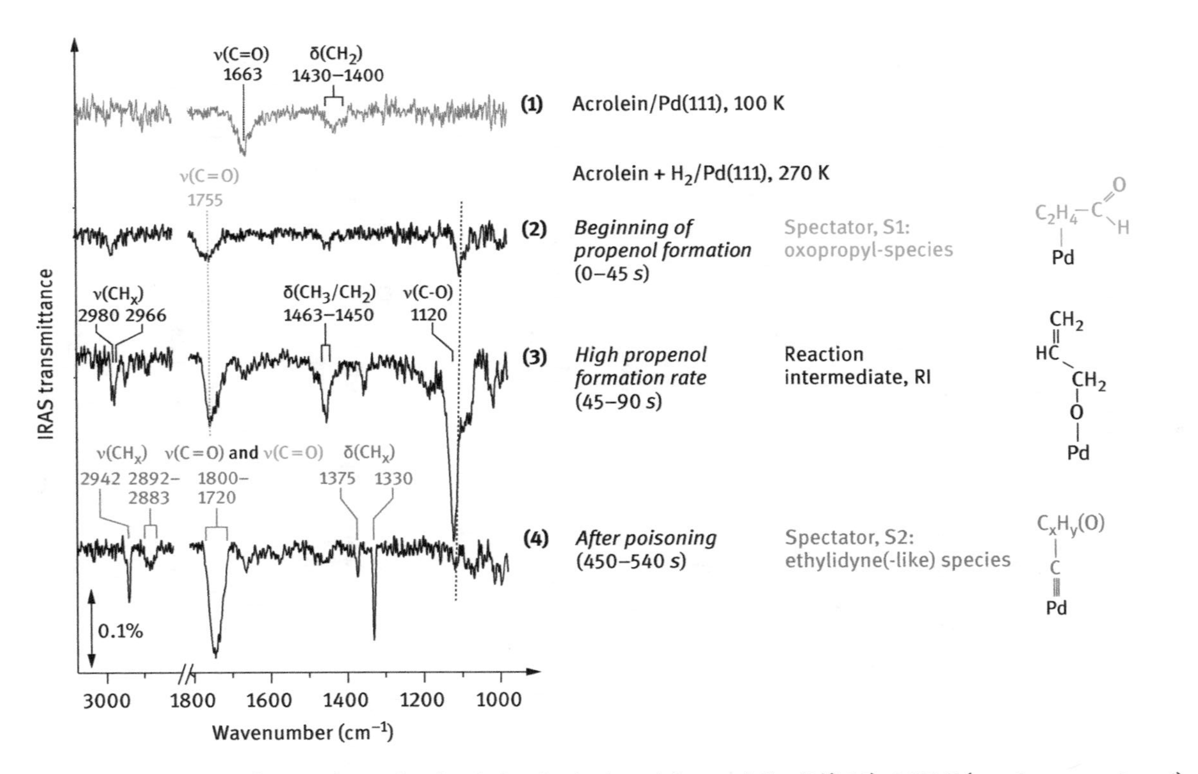

Figure 3.8: IR spectra of a monolayer of molecularly adsorbed acrolein on pristine Pd(1 1 1) at 100 K (gray trace, spectrum 1) and of the surface species formed on Pd(1 1 1) during continuous exposure to acrolein and H_2 at 270 K (black traces, spectra 2–4). The second spectrum 2 was obtained during the induction period and at the beginning of the propenol formation in the gas phase; the spectrum (3) corresponds to the period of high propenol formation rate; the spectrum (4) shows the surface composition of the fully deactivated surface. The feasible structures of identified surface species – both spectators and the reaction intermediate – are shown on the right. The color-coding of the surface species corresponds to the colors of the related vibrational bands.

The band at 1,120 cm^{-1} appears during the first 45 s of acrolein exposure (spectrum (2) in Figure 3.8) and becomes very pronounced in spectrum (3) obtained during the period of the highest propenol formation rates (45–90 s). After the propenol formation rate decreases to zero, the intensity of this band also vanishes (spectrum (4)). Several other bands at 1,090, 1,463–1,450, 2,966, and 2,980 cm^{-1} follow a similar time evolution pattern as shown in Figure 3.8. This behavior shows that the evolution of the surface species associated with the band at 1,120 cm^{-1} is strongly correlated with the formation rate of propenol detected in the gas phase and therefore must be related to a RI.

The intense IR absorption features at 1,090 and 1,120 cm^{-1} are present neither in adsorbed molecular acrolein on Pd nor in acrolein ice and therefore cannot be related to any distinctive vibration of an intact acrolein molecule. The vibration at 1,120 cm^{-1} was previously assigned to a stretching mode of a saturated C–O bond, in which the oxygen atom is coordinated to a metal surface [61, 69, 72–77]. The IR absorption band at 1,090 cm^{-1} can be assigned to a stretching vibration of a saturated C–C bond. In literature, C–C bond vibrations were reported in the range from about 1,000 to 1,130 cm^{-1}, depending on their coordination to the surface [62, 67, 69, 77]. The IR absorption at 1,450–1,463 cm^{-1} appears in a range typical for CH_2 and CH_3 bending vibrations. Tentatively, we assign it to CH_3 asymmetric bending modes, which were reported in the range of 1,450–1,475 cm^{-1} [70, 78, 79]. Alternatively, it could also be related to a CH_2 scissor mode, which typically appears at slightly lower frequency near 1,420–1,430 cm^{-1} [66–69, 80]. The vibrations at 2,966 and 2,980 cm^{-1} can be clearly assigned to C–H stretching vibrations with the band at 2,980 cm^{-1} being related to a C–H bond, in which the C atom is a part of an unsaturated C=C bond [81]. In the region of C=O stretching vibrations, no IR absorption feature can be found that closely follows the evolution of the propenol formation rate. Also, no O–H vibrations can be detected. Based on these spectroscopic signatures, the structure of the RI can be assigned to propenoxy species, which is the result of a half-hydrogenation of acrolein on the C=O double bond. To form this RI, one hydrogen atom attaches to the carbon atom in the C=O bond, thus forming a saturated C–O bond with the vibrational frequency of 1,120 cm^{-1}, in which O is coordinated to Pd atom in an η1-(*O*) configuration (CH_2=CH–CH_2–O–Pd). The C=C double bond is still preserved in the RI as indicated by the C–H stretching frequency of 2,980 cm^{-1} characteristic for a vinyl group. Note that only one additional step – insertion of a second hydrogen atom into the O–Pd bond – is required to form the final reaction product propenol. The proposed structure of the RI is shown in Figure 3.8 on the right. The high intensity of the band at 1,120 cm^{-1}, exceeding even the most intense C=O vibrational band in acrolein and in the oxopropyl species (S1), additionally supports the formation of a C–O bond, which has a large dynamic dipole moment most likely and is expected to have high intensity IR signal.

It is important to emphasize that the desired surface RI propenoxy species is formed not on the pristine Pd(1 1 1) surface, but on the surface covered with a densely packed overlayer of oxopropyl spectator species (S1). By performing sticking coefficient measurement during the induction period, we determined that about one acrolein molecule per four Pd surface atoms was accumulated on Pd(1 1 1) to form the spectator overlayer prior to the onset of propenol formation. Microscopically, this corresponds to a situation that the spectator (S1) species have a formal coverage of 0.25 with respect to the total amount of surface Pd atoms. Most likely, the adsorbed spectator S1 species impose strong geometrical confinement on the adsorption geometry of newly incoming acrolein molecules. In this case, the acrolein adsorption via C=C bond can be suppressed so that acrolein can approach the surface only via the carbonyl group resulting in selective hydrogenation of the C=O bond. Another possible scenario can be based on the dynamic effects, when the potential energy surface of the S1-coverd Pd(1 1 1) steers the incoming acrolein molecules such that they enter the near-surface region only with the O-end. Obviously, the clean Pd(1 1 1) surface is not capable of activating the C=O group toward selective hydrogenation and the strong modification of the surface by S1 is required to trigger the desired selective chemistry.

The last IR spectrum shown in Figure 3.8 was recorded after the propenol formation rate decreased almost to zero (between 450 and 540 s). All features assigned to the half-hydrogenated intermediate RI (propenoxy species) are absent in spectrum (4). Instead, new vibrational bands are observed at 1,330, 1,375 cm^{-1}, in the range 2,883–2,892 cm^{-1}, and at 2,942 cm^{-1}. The sharp peak at 1,330 cm^{-1} is characteristic for the umbrella bending mode of the CH_3 group in ethylidyne or ethylidyne-like species, which were observed in previous studies on Pd(1 1 1) and Pt(1 1 1) [82, 83]. The appearance of these new bands points to partial decomposition of acrolein via decarbonylation reaction yielding a C_2 fragment (e.g., ethylidyne or ethylidine-like species) and probably a fragment containing a carbonyl group, which is in agreement with literature data [46, 84–87]. Since the appearance of this band correlates with deactivation of the catalyst, it can be speculated that the decomposition products block the surface sites and stop the reaction.

3.3.3 Time resolved evolution of the reaction intermediate and the correlation with the formation rate of propenol over Pd(1 1 1)

In order to find detailed correlation between the evolution of the RI on the surface and the appearance of propenol in the gas phase, IRAS studies with higher time resolution have been performed. Figure 3.9a shows a series of IR spectra obtained on the Pd(1 1 1) surface turning over at 270 K. During the acquisition of each spectrum, the surface was exposed to 1.2×10^{13} acrolein molecules.cm^{-2}, which corresponds to 8 s of reaction time displayed in the time axis in Figure 3.9b. Note that after the 6th spectrum, only every fourth spectrum is shown in Figure 3.9a. Approximately in the 2nd or 3rd spectrum, the vibrations related to RI start to appear. The intensities of the peaks grow until about the 7th to 8th spectrum following by disappearing all related features at the end of the reaction.

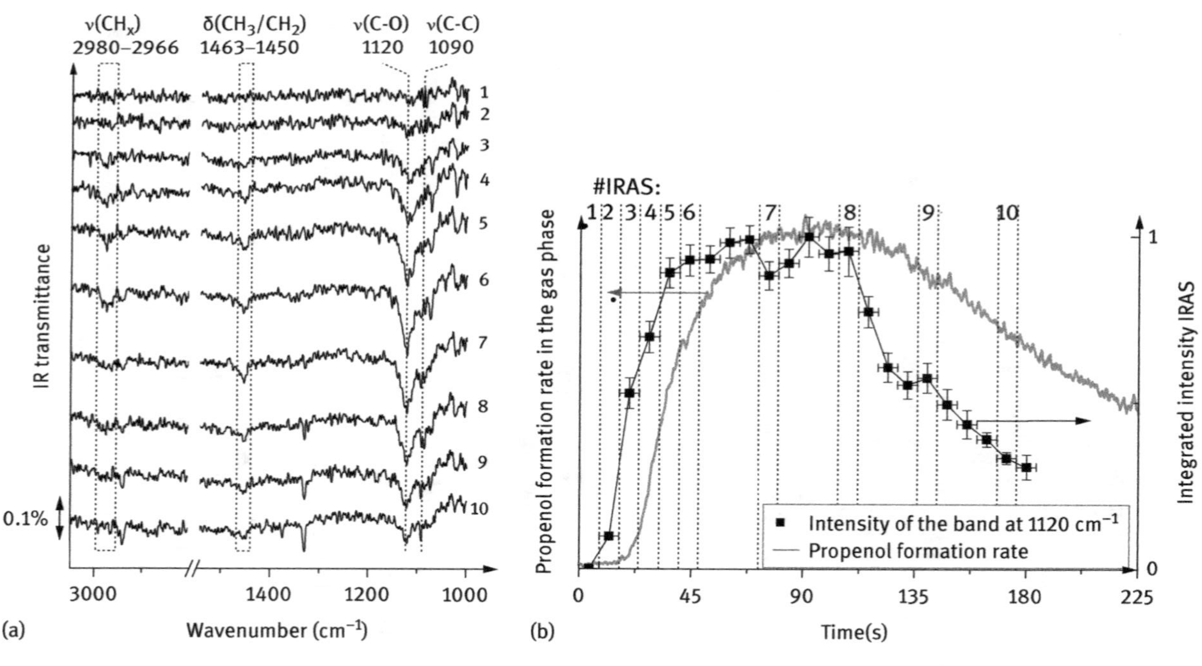

Figure 3.9: The correlation between the formation rate of the reaction intermediate RI on the surface and the evolution of propenol in the gas phase. (a) Series of IR spectra obtained over the Pd(1 1 1) surface turning over at 270 K under continuous exposure to acrolein and H_2. Shown are only the vibrational regions relevant to the reaction intermediate. The spectrum numbers correspond to the reaction times (indicated with dotted lines) that are shown in (b). (b) The integrated intensity of the vibrational peak at 1,120 cm^{-1} (black symbols) and the propenol formation rate in the gas phase (gray line) plotted as a function of reaction time. A clear correlation between the evolution of the reaction intermediate on the surface and propenol in the gas phase is observed.

Figure 3.9b shows the gas-phase formation rate of propenol (gray line) together with the integral intensity of the most intense IR vibration band of the RI at 1,120 cm^{-1}, which can be assumed to approximately reflect the concentration of RI on the surface. It can be clearly seen that the evolution of the propenol formation rate in the gas phase closely follows the concentration of RI on the surface. Thus, the observed strong correlation unambiguously shows that the corresponding propenoxy surface species is a RI that is directly involved in the selective hydrogenation of acrolein to the propenol.

3.3.4 Acrolein hydrogenation over supported Pd nanoparticles

The surface of Pd nanoparticles investigated in this study exhibits ca. 80% of (1 1 1) terraces and 20% of the low-coordinates surface sites such as edges and corners. Despite the large fraction of the (1 1 1) terraces, Pd clusters show the catalytic behavior in acrolein hydrogenation that is strongly different from the extended Pd (1 1 1) surface. In order to understand the origin of the missing catalytic activity toward unsaturated alcohol formation over Pd nanoparticles, we carried out a spectroscopic study on the evolution of surface species under the reaction conditions in the identical way described above for Pd(1 1 1). Prior to the reaction, the surface was continuously exposed to hydrogen and then the acrolein molecular beam was switched on at the time moment indicated as zero. Figure 3.10a shows the evolution of both reaction products – propanal and propenol – during the reaction. Consistent with the data shown in Figure 3.6a, c, no formation of propenol was observed on this surface and only minor formation rate of propanal could be detected, which passes through a small maximum and returns to zero. The evolution of the surface species formed in the course of the reactions was monitored by IRAS. The IR spectra corresponding to regions 1–4, which are indicated on the kinetic curve of Figure 3.10a, are shown in Figure 3.10b. Obviously, a completely different composition of surface adsorbates is formed on Pd particles compared to that on Pd(1 1 1). The spectra are dominated by the features in the range 1,800–1,960 cm^{-1}, which appear first in the lower wavenumbers and grow in intensity and red shift with increasing reaction time. These vibrational signatures can be clearly related to the accumulation of CO molecules on the surface, which result from acrolein decarbonylation. Previously, very similar evolution of vibrational bands was observed for consecutive adsorption of CO molecules on Pd particles in a similar size range [88]. The hypothesis on CO formation from acrolein decomposition is in agreement with numerous literature reports showing that acrolein and the higher α,β-unsaturated ketones and aldehydes can readily undergo decarbonylation over transition metal surfaces [46, 52]. The obtained infrared data strongly suggest that acrolein undergoes decarbonylation on Pd nanoparticles forming CO molecules that block the surface and most likely prevent formation of closely packed spectator (S1) overlayers required for selective acrolein hydrogenation

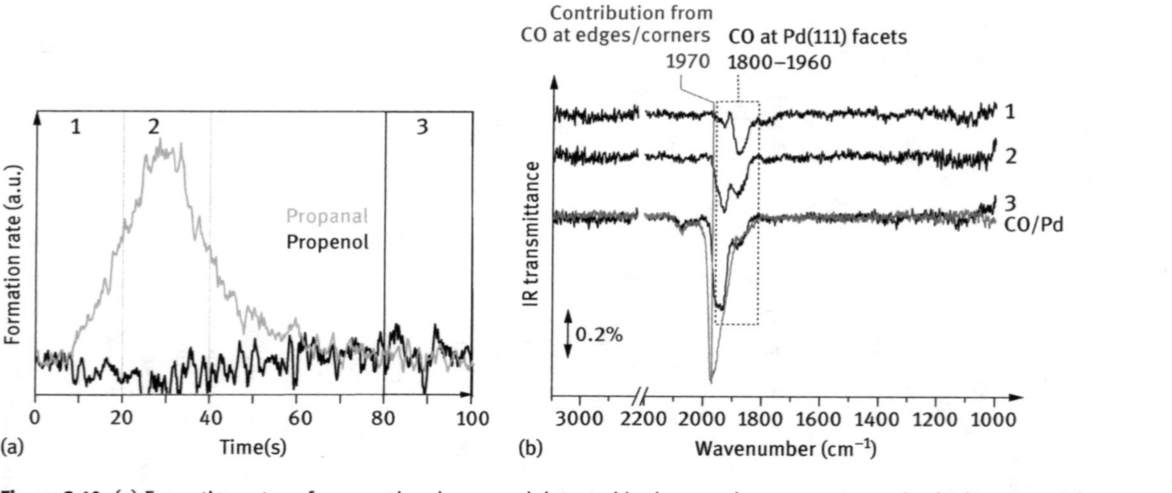

Figure 3.10: (a) Formation rates of propanal and propenol detected in the gas phase over 12-nm-sized Pd nanoparticles supported on Fe_3O_4/Pt(1 1 1) model film at 270 K and (b) simultaneously recorded IR spectra on this surface. The time resolution of the spectra is 20 s; the spectra 1–3 correspond to the regions 1–3 indicated in (a), after Ref. [24].

to unsaturated alcohol. Since the Pd clusters are mainly terminated by (1 1 1) facets [30], which are not active in acrolein decarbonylation in this temperature range as observed on Pd(1 1 1), most likely the low-coordinated surface sites of Pd nanoparticles (edges, corners, (1 0 0) facets) are responsible for the facile acrolein decomposition and formation of CO. It should be noted that not only alcohol formation but also hydrogenation of the C=C bond, which is discussed to be generally facile even over carbon-containing surfaces, is prevented on the surface covered by CO.

The final mechanistic model of partial selective hydrogenation of acrolein over two investigated surfaces is shown in Figure 3.11. The desired reaction pathway – selective hydrogenation of the C=O bond in acrolein – was observed over Pd(1 1 1) surface with nearly 100% selectivity, while over oxide supported Pd nanoparticles only C=C bond hydrogenation to form minor amounts of propanal occurred. The selectivity in hydrogenation of the C=O bond critically depends on the presence of oxopropyl spectator species, which form a dense overlayer with the surface coverage close to 0.25 during the induction period of the reaction (Figure 3.11a). These spectators result from the addition of one H atom to the C=C bond of acrolein to form oxopropyl. After the spectator overlayer is formed, acrolein can adsorb on this modified surface only via the C=O bond, which is first hydrogenated to a propenoxy RI attached via the O-end to Pd and then to unsaturated alcohol. The nature of the surface RI was established spectroscopically.

During the course of the reaction, the simultaneous evolution of the propenoxy reactive intermediates on the surface and formation of the propenol product in the gas phase was monitored by IRAS and QMS, respectively. With this, a direct assignment of one of the surface species to a RI was achieved, while the other surface species (oxopropyl and ethylidyne) were identified as spectators. On supported Pd nanoparticles, formation of a spectator overlayer was found to be prevented by strong acrolein decarbonylation and the surface was observed to be active only for hydrogenation of the C=C bond.

Obtained atomistic-level insights into chemoselective hydrogenation chemistry of acrolein highlight the exceptional importance of spectator species which are usually formed on the catalytically active surface under reaction conditions. Related effects are expected to play a key role in controlling chemoselectivity in hydrogenation of all types of α,β-unsaturated aldehydes and ketones and hold a great potential for further development of new selective powdered catalysts, such as ligand-modified nanoparticles.

3.4 Conclusions

In this chapter, we have compiled a range of the latest experimental studies to address some of the important questions that needs to be understood if aiming at

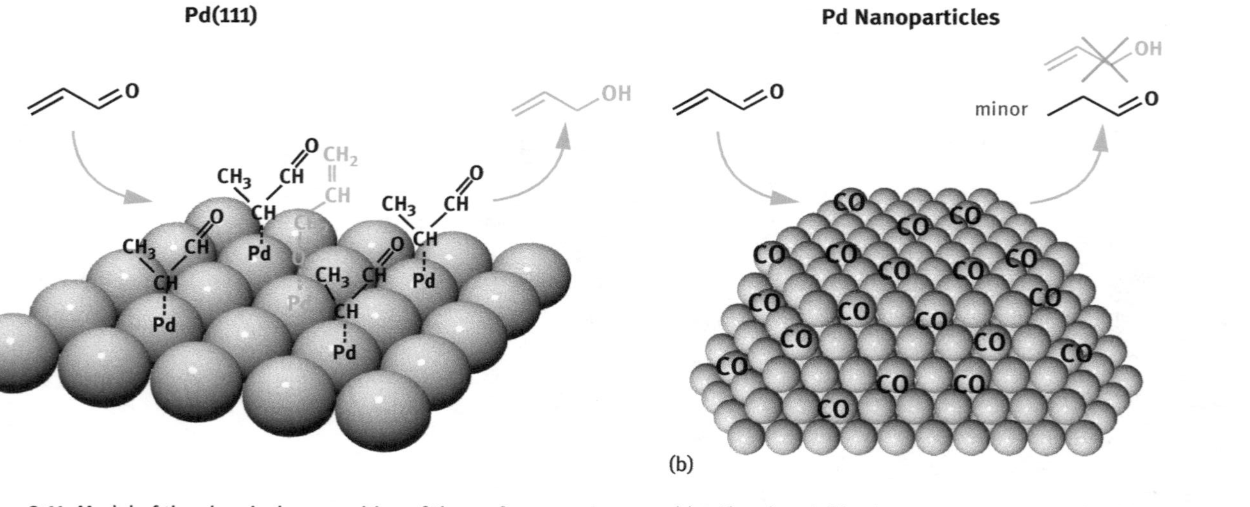

Figure 3.11: Model of the chemical composition of the surfaces turning over (a) Pd(1 1 1) and (b) Pd nanoparticles. On Pd(1 1 1), a dense overlayer of spectator (oxopropyl) species is formed at the initial stages of reaction that allows activation of C=O double bond and renders the surface chemoselective toward

an atomistic understanding of heterogeneously catalyzed hydrogenation over transition metal surfaces. In these studies, well-defined nanostructured supported model catalysts were employed, which are based on single-crystalline oxide films epitaxially grown on single-crystalline metals. These systems were proven to be suitable models to address the catalytic properties of heterogeneous oxide–base catalysts. Complementary, the reactions were investigated also on the extended single-crystalline metals to explore how the nanoscopic nature of small metal clusters affects their catalytic activity and the mechanisms of the underlying surface processes. The basis for the atomistic understanding is a detailed knowledge of the structural properties of the catalytic systems, which can be correlated to their catalytic performance and the reaction mechanisms with the rigor of modern surface science methodology.

Using two selected catalytic systems, we have shown how the machinery of modern surface science can be used to explore the reaction mechanisms and the details of the reaction kinetics. First, we provide an example on hydrogenation of simple olefins over Pd catalysts and show that rather complex reaction mechanisms including not only surface but also subsurface species constitute the basis of the observed kinetics. Particularly, subsurface hydrogen plays a crucial role in hydrogenation of simple olefins over Pd and the rate, at which the concentration of subsurface hydrogen can be replenished, limits the overall reaction rate. This observation clearly demonstrates that the permeability of a catalytic surface to hydrogen and the structural factors governing this property (such as lateral flexibility of nanoparticles) can be of crucial importance for the overall catalytic performance.

In our second example, the selective particle hydrogenation of α,β-unsaturated aldehydes and ketones was investigated by combination infrared spectroscopy and molecular beam techniques. We observed an exceptionally high selectivity toward the desired reaction product unsaturated alcohol on Pd(1 1 1) single crystal – the result, which cannot be achieved on the real powdered catalysts. By monitoring surface composition during the reaction, we found that the exceptionally high selectivity is due to formation of an overlayer of spectator species, which most likely impose a geometrical constrain on the adsorption of reactants and render the surface highly selective toward the desired reaction route. The concept of the co-adsorbed spectator species can be further extended to rational design of surfaces with the desired catalytic properties by formation of ligand overlayer, which govern the adsorption geometry of the reactants and with this finely tune selectivity.

References

[1] Bond GC. Introduction to the Catalysis of Hydrocarbon Reactions. Metal-Catalysed Reactions of Hydrocarbons. Boston, MA: Springer US; 2005: 209–55.
[2] Somorjai GA. Introduction to Surface Chemistry and Catalysis, 2nd edn. New York: John Wiley & Sons; 2010.

[3] Kemball C. The Catalytic Exchange of Hydrocarbons with Deuterium. Adv Cat 1959; 11: 223–62.
[4] Zaera F. Probing Catalytic Reactions at Surfaces. Prog Surf Sci 2001; 69: 1–98.
[5] Schauermann S, Nilius N, Shaikhutdinov S, Freund H-J. Nanoparticles for Heterogeneous Catalysis: New Mechanistic Insights. Acc Chem Res 2012; 46: 1673–81.
[6] Schauermann S, Freund H-J. Model Approach in Heterogeneous Catalysis: Kinetics and Thermodynamics of Surface Reactions. Acc Chem Res 2015; 48: 2775–82.
[7] Schauermann S, Silbaugh TL, Campbell CT. Single-Crystal Adsorption Calorimetry on Well-Defined Surfaces: From Single Crystals to Supported Nanoparticles. Chem Rec 2014; 14: 759–74.
[8] Bäumer M, Freund H-J. Metal Deposits on Well-Ordered Oxide Films. Prog Surf Sci 1999; 61: 127–98.
[9] Henry CR. Surface Studies of Supported Model Catalysts. Surf Sci Rep 1998; 31: 231–326.
[10] Goodman DW. Model Catalysts – From Extended Single Crystals to Supported Particles. Surf Rev Lett 1995; 2: 9–24.
[11] Campbell CT. Ultrathin Metal Films and Particles on Oxide Surfaces: Structural, Electronic and Chemisorptive Properties. Surf Sci Rep 1997; 27: 1–111.
[12] Peter M, Flores Camacho JM, Adamovski S, et al. Trends in the Binding Strength of Surface Species on Nanoparticles: How Does the Adsorption Energy Scale with the Particle Size? Angew Chem Int Ed 2013; 52: 5175–9.
[13] Dementyev P, Dostert K-H, Ivars-Barceló F, et al. Water Interaction with Iron Oxides. Angew Chem Int Ed 2015; 54: 13942–6.
[14] Wilde M, Fukutani K, Ludwig W, et al. Influence of Carbon Deposition on the Hydrogen Distribution in Pd Nanoparticles and Their Reactivity in Olefin Hydrogenation. Angew Chem Int Ed 2008; 47: 9289–93.
[15] Horiuti J, Miyahara K. Hydrogenation of Ethylene on Metallic Catalysts. Washington, DC: National Bureau of Standards; 1968.
[16] Doyle AM, Shaikhutdinov SK, Jackson SD, Freund H-J. Hydrogenation on Metal Surfaces: Why are Nanoparticles More Active than Single Crystals? Angew Chem Int Ed 2003; 42: 5240–3.
[17] Lee I, Zaera F. Thermal Chemistry of C4 Hydrocarbons on Pt(111): Mechanism for Double-Bond Isomerization. J Phys Chem B 2005; 109: 2745–53.
[18] Brandt B, Fischer J-H, Ludwig W, et al. Isomerization and Hydrogenation of cis-2-Butene on Pd Model Catalyst. J Phys Chem C 2008; 112: 11408–20.
[19] Ludwig W, Savara A, Brandt B, Schauermann S. A Kinetic Study on the Conversion of Cis-2-butene with Deuterium on a Pd/Fe3O4 Model Catalyst. Phys Chem Chem Phys 2011; 13: 966–77.
[20] Ludwig W, Savara A, Dostert K-H, Schauermann S. Olefin Hydrogenation on Pd Model Supported Catalysts: New Mechanistic Insights. J Catal 2011; 284: 148–56.
[21] Ludwig W, Savara A, Madix RJ, Schauermann S, Freund H-J. Subsurface Hydrogen Diffusion into Pd Nanoparticles: Role of Low-Coordinated Surface Sites and Facilitation by Carbon. J Phys Chem C 2012; 116: 3539–44.
[22] Ludwig W, Savara A, Schauermann S, Freund H-J. Role of Low-Coordinated Surface Sites in Olefin Hydrogenation: A Molecular Beam Study on Pd Nanoparticles and Pd(111). Chem Phys Chem 2010; 11: 2319–22.
[23] Ludwig W, Savara A, Schauermann S. Role of Hydrogen in Olefin Isomerization and Hydrogenation: A Molecular Beam Study on Pd Model Supported Catalysts. Dalton Trans 2010; 39: 8484–91.
[24] Dostert K-H, O'Brien CP, Ivars-Barceló F, Schauermann S, Freund H-J. Spectators Control Selectivity in Surface Chemistry: Acrolein Partial Hydrogenation Over Pd. J Amer Chem Soc 2015; 137: 13496–502.

[25] K.-H. Dostert, C. P. O'Brien, W. Liu, W. Riedel, A. Savara, A. Tkatchenko, S. Schauermann, H.-J. Freund, Surf. Sci. 2016, 650, 149–160.
[26] Dostert K-H, O'Brien CP, Mirabella F, Ivars-Barcelo F, Schauermann S. Adsorption of Acrolein, Propanal, and Allyl alcohol on Pd(111): A Combined Infrared Reflection-Absorption Spectroscopy and Temperature Programmed Desorption Study. Phys Chem Chem Phys 2016; 18: 13960–73.
[27] C. P. O'Brien, K. H. Dostert, M. Hollerer, C. Stiehler, F. Calaza, S. Schauermann, S. Shaikhutdinov, M. Sterrer, H. J. Freund, Faraday Discuss. 2016, 188, 309–321.
[28] O'Brien CP, Dostert K-H, Schauermann S, Freund H-J. Selective Hydrogenation of Acrolein Over Pd Model Catalysts: Temperature and Particle-Size Effects. Chem A Eur J 2016; 22: 15856–63.
[29] Neyman KM, Schauermann S. Hydrogen Diffusion into Palladium Nanoparticles: Pivotal Promotion by Carbon. Angew Chem Int Ed 2010; 49: 4743–6.
[30] Schalow T, Brandt B, Starr DE, et al. Particle Size Dependent Adsorption and Reaction Kinetics on Reduced and Partially Oxidized Pd Nanoparticles. Phys Chem Chem Phys 2007; 9: 1347–61.
[31] Wilde M, Fukutani K, Naschitzki M, Freund HJ. Hydrogen Absorption in Oxide-Supported Palladium Nanocrystals. Phys Rev B 2008; 77: 113412.
[32] Aleksandrov HA, Kozlov SM, Schauermann S, Vayssilov GN, Neyman KM. How Absorbed Hydrogen Affects the Catalytic Activity of Transition Metals. Angew Chem Int Ed 2014; 53: 13371–5.
[33] Aleksandrov HA, Viñes F, Ludwig W, Schauermann S, Neyman KM. Tuning the Surface Chemistry of Pd by Atomic C and H: A Microscopic Picture. Chem A Eur J 2013; 19: 1335–45.
[34] Horiuti I, Polanyi M. Exchange Reactions of Hydrogen on Metallic Catalysts. Trans Faraday Soc 1934; 30: 1164–72.
[35] Zaera F. An Organometallic Guide to the Chemistry of Hydrocarbon Moieties on Transition Metal Surfaces. Chem Rev 1995; 95: 2651–93.
[36] S. P. Daley, A. L. Utz, T. R. Trautman, S. T. Ceyer, J. Amer. Chem. Soc. 1994, 116, 6001–6002
[37] Teschner D, Borsodi J, Wootsch A, et al. The Roles of Subsurface Carbon and Hydrogen in Palladium-Catalyzed Alkyne Hydrogenation. Science 2008; 320: 86–9.
[38] Schalow T, Brandt B, Starr D, et al. Oxygen-Induced Restructuring of a Pd/Fe_3O_4 Model Catalyst. Catal Lett 2006; 107: 189–96.
[39] Schauermann S, Hoffmann J, Johánek V, Hartmann J, Libuda J, Freund H-J. Catalytic Activity and Poisoning of Specific Sites on Supported Metal Nanoparticles. Angew Chem Int Ed 2002; 41: 2532–5.
[40] Christmann K. Interaction of Hydrogen with Solid Surfaces. Surf Sci Rep 1988; 9: 1–163.
[41] Jewell LL, Davis BH. Review of Absorption and Adsorption in the Hydrogen–Palladium System. Appl Catal A Gen 2006; 310: 1–15.
[42] Savara A, Ludwig W, Schauermann S. Kinetic Evidence for a Non-Langmuir-Hinshelwood Surface Reaction: H/D Exchange Over Pd Nanoparticles and Pd(111). Chem Phys Chem 2013; 14: 1686–95.
[43] Gallezot P, Richard D. Selective Hydrogenation of Alpha, Beta-Unsaturated Aldehydes. Catal Rev-Sci Eng 1998; 40: 81–126.
[44] Mäki-Arvela P, Hájek J, Salmi T, Murzin DY. Chemoselective Hydrogenation of Carbonyl Compounds Over Heterogeneous Catalysts. Appl Catal A Gen 2005; 292: 1–49.
[45] Claus P. Selective Hydrogenation of ά,β-Unsaturated Aldehydes and Other C=O and C=C Bonds Containing Compounds. Top Catal 1998; 5: 51–62.
[46] de Jesús JC, Zaera F. Adsorption and Thermal Chemistry of Acrolein and Crotonaldehyde on Pt (111) Surfaces. Surf Sci 1999; 430: 99–115.
[47] Brandt K, Chiu ME, Watson DJ, Tikhov MS, Lambert RM. Chemoselective Catalytic Hydrogenation of Acrolein on Ag(111): Effect of Molecular Orientation on Reaction Selectivity. J Am Chem Soc 2009; 131: 17286–90.

[48] Wei H, Gomez C, Liu J, et al. Selective Hydrogenation of Acrolein on Supported Silver Catalysts: A Kinetics Study of Particle Size Effects. J Catal 2013; 298: 18–26.

[49] Bron M, Teschner D, Knop-Gericke A, et al. Bridging the Pressure and Materials Gap: In-Depth Characterisation and Reaction Studies of Silver-Catalysed Acrolein Hydrogenation. J Catal 2005; 234: 37–47.

[50] Bron M, Teschner D, Knop-Gericke A, et al. Silver as Acrolein Hydrogenation Catalyst: Intricate Effects of catalyst Nature and Reactant Partial Pressures. Phys Chem Chem Phys 2007; 9: 3559–69.

[51] Kliewer CJ, Bieri M, Somorjai GA. Hydrogenation of the α,β-Unsaturated Aldehydes Acrolein, Crotonaldehyde, and Prenal over Pt Single Crystals: A Kinetic and Sum-Frequency Generation Vibrational Spectroscopy Study. J Am Chem Soc 2009; 131: 9958–66.

[52] Murillo LE, Chen JG. A comparative study of the adsorption and hydrogenation of acrolein on Pt(111), Ni(111) film and Pt–Ni–Pt(111) bimetallic surfaces. Surf Sci 2008; 602: 919–31.

[53] Birchem T, Pradier CM, Berthier Y, Cordier G. Hydrogenation of 3-Methyl-Crotonaldehyde on the Pt(553) Stepped Surface: Influence of the Structure and of Preadsorbed Tin. J Catal 1996; 161: 68–77.

[54] Englisch M, Ranade VS, Lercher JA. Hydrogenation of Crotonaldehyde over Pt Based Bimetallic Catalysts. J Mol Catal A Chem 1997; 121: 69–80.

[55] Marinelli TBLW, Nabuurs S, Ponec V. Activity and Selectivity in the Reactions of Substituted α,β-Unsaturated Aldehydes. J Catal 1995; 151: 431–8.

[56] Hutchings GJ, King F, Okoye IP, Padley MB, Rochester CH. Selectivity Enhancement in the Hydrogenation of α, β-Unsaturated Aldehydes and Ketones Using Thiophene-Modified Catalysts. J Catal 1994; 148: 453–63.

[57] Pradier CM, Birchem T, Berthier Y, Cordier G. Hydrogenation of 3-Methyl-Butenal on Pt(110); Comparison with Pt(111). Catal Lett 1994; 29: 371–8.

[58] Vannice MA. The Influence of MSI (metal–support interactions) on Activity and Selectivity in the Hydrogenation of Aldehydes and Ketones. Top Catal 1997; 4: 241–8.

[59] Englisch M, Jentys A, Lercher JA. Structure Sensitivity of the Hydrogenation of Crotonaldehyde over Pt/SiO2and Pt/TiO2. J Catal 1997; 166: 25–35.

[60] Kennedy G, Baker LR, Somorjai GA. Selective Amplification of CJO Bond Hydrogenation on Pt/TiO2: Catalytic Reaction and Sum-Frequency Generation Vibrational Spectroscopy Studies of Crotonaldehyde Hydrogenation. Angew Chem Int Ed 2014; 53: 3405–8.

[61] Haubrich J, Loffreda D, Delbecq F, et al. Adsorption and Vibrations of α,β-Unsaturated Aldehydes on Pure Pt and Pt–Sn Alloy (111) Surfaces I. Prenal. J Phys Chem C 2008; 112: 3701–18.

[62] Haubrich J, Loffreda D, Delbecq F, et al. Adsorption of α,β-Unsaturated Aldehydes on Pt(111) and Pt–Sn Alloys: II. Crotonaldehyde. J Phys Chem C 2009; 113: 13947–67.

[63] Loffreda D, Delbecq F, Vigné F, Sautet P. Catalytic Hydrogenation of Unsaturated Aldehydes on Pt(111): Understanding the Selectivity from First-Principles Calculations. Angew Chem Int Ed 2005; 44: 5279–82.

[64] Ide MS, Hao B, Neurock M, Davis RJ. Mechanistic Insights on the Hydrogenation of α,β-Unsaturated Ketones and Aldehydes to Unsaturated Alcohols over Metal Catalysts. ACS Catal 2012; 2: 671–83.

[65] Ponec V. On the Role of Promoters in Hydrogenations on Metals; α,β-Unsaturated Aldehydes and Ketones. Appl Catal A Gen 1997; 149: 27–48.

[66] Akita M, Osaka N, Itoh K. Infra-Red Reflection Absorption Spectroscopic Study on Adsorption Structures of Acrolein on Polycrystalline Gold and Au(111) Surfaces under ultra-High Vacuum Conditions. Surf Sci 1998; 405: 172–81.

[67] Fujii S, Osaka N, Akita M, Itoh K. Infrared Reflection Absorption Spectroscopic Study on the Adsorption Structures of Acrolein on an Evaporated Silver Film. J Phys Chem 1995; 99: 6994–7001.
[68] Hamada Y, Nishimura Y, Tsuboi M. Infrared Spectrum of Trans-Acrolein. Chem Phys 1985; 100: 365–375.
[69] Loffreda D, Jugnet Y, Delbecq F, Bertolini JC, Sautet P. Coverage Dependent Adsorption of Acrolein on Pt(111) from a Combination of First Principle Theory and HREELS Study. J Phys Chem B 2004; 108: 9085–93.
[70] Colthup NB, Daly LH, Wiberley SE. CHAPTER 9 – CARBONYL COMPOUNDS. In: Colthup NB, Daly LH, Wiberley SE, eds. Introduction to Infrared and Raman Spectroscopy, 3rd edn. San Diego: Academic Press; 1990: 289–325.
[71] Mecke B, Noack K. Untersuchungen über die Beeinflussung der Frequenz und Intensität der νC=O- und νC=C-Banden im I.R.-Spektrum ungesättigter Ketone durch Konjugation und sterische Hinderung. Spectrochimica Acta 1958; 12: 391–3.
[72] Mitchell WJ, Xie J, Jachimowski TA, Weinberg WH. Carbon Monoxide Hydrogenation on the Ru (001) Surface at Low Temperature Using Gas-Phase Atomic Hydrogen: Spectroscopic Evidence for the Carbonyl Insertion Mechanism on a Transition Metal Surface. J Am Chem Soc 1995; 117: 2606–17.
[73] Weldon MK, Friend CM. Probing Surface Reaction Mechanisms Using Chemical and Vibrational Methods: Alkyl Oxidation and Reactivity of Alcohols on Transitions Metal Surfaces. Chem Rev 1996; 96: 1391–412.
[74] Borasio M, Rodríguez de la Fuente O, Rupprechter G, Freund H-J. In Situ Studies of Methanol Decomposition and Oxidation on Pd(111) by PM-IRAS and XPS Spectroscopy. J Phys Chem B 2005; 109: 17791–4.
[75] Schauermann S, Hoffmann J, Johanek V, Hartmann J, Libuda J. Adsorption, Decomposition and Oxidation of Methanol on Alumina Supported Palladium Particles. Phys Chem Chem Phys 2002; 4: 3909–18.
[76] Sexton BA. Methanol Decomposition on Platinum (111). Surf Sci 1981; 102: 271–81.
[77] Davis JL, Barteau MA. Polymerization and Decarbonylation Reactions of Aldehydes on the Pd (111) Surface. J Am Chem Soc 1989; 111: 1782–92.
[78] Colthup NB, Daly LH, Wiberley SE. CHAPTER 5 – METHYL AND METHYLENE GROUPS. In: Colthup NB, Daly LH, Wiberley SE, eds. Introduction to Infrared and Raman Spectroscopy, 3rd edn. San Diego: Academic Press; 1990: 215–33.
[79] Guirgis GA, Drew BR, Gounev TK, Durig JR. Conformational Stability and Vibrational Assignment of Propanal. Spectrochimica Acta Part A: Mol Biomol Spectrosc 1998; 54: 123–43.
[80] Puzzarini C, Penocchio E, Biczysko M, Barone V. Molecular Structure and Spectroscopic Signatures of Acrolein: Theory Meets Experiment. J Phys Chem A 2014; 118: 6648–56.
[81] Colthup NB, Daly LH, Wiberley SE. CHAPTER 7 - OLEFIN GROUPS. In: Colthup NB, Daly LH, Wiberley SE, eds. Introduction to Infrared and Raman Spectroscopy, 3rd edn. San Diego: Academic Press; 1990: 247–60.
[82] Hill JM, Shen J, Watwe RM, Dumesic JA. Microcalorimetric, Infrared Spectroscopic, and DFT Studies of Ethylene Adsorption on Pd and Pd/Sn Catalysts. Langmuir 2000; 16: 2213–9.
[83] Mohsin SB, Trenary M, Robota HJ. Kinetics of Ethylidyne Formation on Pt(111) From Time-Dependent Infrared Spectroscopy. Chem Phys Lett 1989; 154: 511–5.
[84] Brown NF, Barteau MA. Reactions of Unsaturated Oxygenates on Rhodium(111) as Probes of Multiple Coordination of Adsorbates. J Am Chem Soc 1992; 114: 4258–65.
[85] de Jesús JC, Zaera F. Double-Bond Activation in Unsaturated Aldehydes: Conversion of Acrolein to Propene and Ketene on Pt(111) Surfaces. J Mol Catal A Chem 1999; 138: 237–40.

[86] Davis JL, Barteau MA. Vinyl Substituent Effects on the Reactions of Higher Oxygenates on Pd (111). J Mol Catal 1992; 77: 109–24.
[87] Haubrich J, Loffreda D, Delbecq F, et al. Mechanistic and Spectroscopic Identification of Initial Reaction Intermediates for Prenal Decomposition on a Platinum Model Catalyst. Phys Chem Chem Phys 2011; 13: 6000–9.
[88] Sandell A, Libuda J, Brühwiler PA, et al. Interaction of CO with Pd Clusters Supported on a Thin Alumina Film. J Vac Sci Technol A 1996; 14: 1546–51.

Ahmed K. A. AlAsseel, S. David Jackson and Kathleen Kirkwood

4 Aromatic hydrogenation

4.1 Introduction

In 1901, Sabatier and Senderence [1] reported the first catalytic aromatic hydrogenation: they "attacked" the benzene ring with hydrogen at atmospheric pressure and temperatures between 343 K and 473 K over a nickel catalyst and succeeded in converting it to cyclohexane. This was the first example of hydrogenation of an aromatic ring. Nowadays, aromatic hydrogenation is a major industrial process with around 4.6 mT of benzene hydrogenated to cyclohexane each year. There have been many reviews on this topic over the years. One of the most extensive reviews was that of Stanislaus and Cooper in 1994 [2] with a more up-to-date review of aromatic hydrogenation by Bond in 2005 [3] where he focused on the hydrogenation of the aromatic ring in hydrocarbons. This chapter intends to build upon Bond's review by further discussing hydrocarbon ring hydrogenation and then introducing oxygen and nitrogen-substituted aromatics before examining hydrogenation of fused aromatic rings.

The aromatic ring is highly stable and has an unusually large resonance energy. Benzene, the simplest aromatic, contains no distinct single or double bonds; instead, there is delocalization to form the stabilizing electron clouds above and below the aromatic ring, with all C–C bonds in benzene having an identical bond length of 1.40 Å. Even going back as far as 1936, researchers have investigated the stabilization energy of the benzene ring. Kistiakowsky carried out experimental work studying the heats of hydrogenation of cyclohexene, 1,3-cyclohexadiene, and benzene to determine the stabilizing effect of the aromatic ring [4, 5]. This work obtained heats of hydrogenation of –119.7 $kJ.mol^{-1}$, 231.8 $kJ.mol^{-1}$, and 208.5 $kJ.mol^{-1}$ for cyclohexene, 1,3-cyclohexadiene, and benzene, respectively, revealing that the hydrogenation of benzene to 1,3-cyclohexadiene would be an endothermic reaction. Indeed, the free energy change for the hydrogenation of benzene to 1,3-cyclohexadiene is positive. The overall resonance stabilization energy for benzene can be easily shown to be 149.5 $kJ.mol^{-1}$.

The addition of substituents to the aromatic ring brings changes in electron density, which can change the strength of adsorption and hence reactivity. Substituents can also have steric effects and may adsorb in competition to the aromatic ring. These aspects will be more thoroughly examined later in the chapter as the hydrogenation of substituted benzenes is discussed.

4.1.1 Importance of aromatic hydrogenation

The simplest aromatic known is benzene, and in 2015, the global demand for the chemical use was estimated to be 46 million tons. It is one of the seven key organic

https://doi.org/10.1515/9783110545210-004

building blocks in today's petrochemical industry and is used as the starting point for a wide variety of products including ethylbenzene, the precursor to styrene; nitrobenzene, the precursor to aniline; and cumene, the precursor to phenol. The other major product accounting for around 11% of benzene usage is cyclohexane. The demand for cyclohexane is linked with the production of nylon fibers, both Nylon 6 and Nylon 6,6. The reaction scheme for this process is shown in Figure 4.1.

+ $3H_2$ ⟶

Figure 4.1: Hydrogenation of benzene.

The hydrogenation of benzene is a highly exothermic process (ΔH ~−206 kJ.mol^{-1}); so, managing the process heat balance is key factor. A large number of gas-phase and liquid-phase processes have been developed, e.g., Process HA-84 (Engelhard Industries & Sinclair Research Inc.), the Hydrar Process (U.O.P. part of the Honeywell group), and the I.F.P. Process (Institut Francais du Petrole), among others. Typical catalysts are based on platinum or, more commonly, nickel; for example, the IFP liquid-phase process uses Raney Nickel™. The liquid-phase processes operate at medium pressure (20–50 bar) and temperature ($<$ 250°C), whereas the gas-phase processes operate at higher temperatures ($>$ 400°C) but similar pressures. All of the processes use hydrogen in excess, which is recycled, to achieve a high conversion ($>$ 99%) and carefully avoid isomerization to methylcyclopentane either by using low temperatures or high space velocities.

Academically, benzene hydrogenation has been used in many cases as a model reaction to test catalytic activity; however, as will be discussed later, it may be that this is a poor choice. The reactions are normally carried out in the liquid phase using pressurized systems or in the gas phase using flow-through reactors. Most work has dominated on single-metal supported systems, in particular Pt, Pd, Rh, and Ni as well as further work on bi- and multimetallic systems [6]. However, other factors such as dispersion, metal crystallite size, and the support can all influence catalyst activity.

4.1.2 Early work on benzene hydrogenation

The reversible reaction between benzene, hydrogen, and cyclohexane has been the subject of many investigations since the original findings of Sabatier and Senderens in 1901 [1]. Their work established the use of nickel-based hydrogenation of unsaturated molecules to their saturated equivalent including the reaction of benzene to cyclohexane, for which Sabatier was awarded a Nobel Prize in 1912. The hydrogenation was carried out by passing hot vapor of the organic molecule and hydrogen over hot, finely divided nickel at 200°C. For the hydrogenation of benzene to cyclohexane, temperatures in the range 70–190°C were found to be effective with rate reaching a maximum between 170°C and 190°C [7]. Note that at this temperature, the hydrogenation is still favored thermodynamically with a ΔG at 227°C of −21.6 kJ.mol^{-1}. If the temperature is increased above 300°C, then the further reduction of benzene to methane occurs and carbon deposits develop on the nickel surface, which can lead to deactivation.

In 1967, the kinetics of benzene hydrogenation over cobalt-supported catalysts was investigated by Taylor and Staffin in a differential flow reactor [8]. From their work, it was found that at low temperature the rate of benzene exhibited near zero-order dependence with respect to benzene and first order in hydrogen, with increased temperature increasing both the orders in hydrogen and benzene. With many researchers proposing a mechanism which involves π-bonded intermediates, this low order in benzene is in agreement with a strongly bound substrate to the metal surface. These results are consistent with the mechanism proposed by many researchers on the hydrogenation of benzene, where a strongly bound benzene molecule is hydrogenated with a kinetically slow step related to the loss of aromatic character, followed by rapid hydrogenation to cyclohexane via cyclohexene.

Although the proposed mechanism of the hydrogenation of benzene has gathered considerable interest over the years, a generally accepted mechanism is still lacking. Early work by Chou and Vannice in 1987 studied the hydrogenation of benzene over palladium and believed the Langmuir–Hinshelwood (L–H) model could not be considered as the exact mechanism of action because there is no single rate-determining step and indeed mechanisms invoking both L–H kinetics and Rideal–Eley kinetics have been proposed [9–11]. One reason for this ambiguity relates to the adsorption of hydrogen as it has been suggested by many researchers that the hydrogen involved in the hydrogenation reaction is weakly adsorbed on the surface of the metal. Indeed, much of the work on benzene is contradictory and yet the quality of the individual studies is high, suggesting that the hydrogenation of benzene is complex and may be catalyst specific.

4.1.3 Recent work on benzene hydrogenation

As indicated above, benzene hydrogenation has received substantial attention throughout the development of catalytic hydrogenation, due in no small part to its industrial relevance and this attention continues to the present day. Nickel-supported catalysts are currently employed industrially to carry out the reaction and so researchers have focused on nickel–alumina and in particular utilizing different preparation methods to achieve high activity. These methods included deposition of nickel on the high surface area alumina, nickel–alumina coprecipitation and sol–gel. The latter has gained substantial interest as it is believed to result in a higher activity catalyst due to an optimum match of dispersion and support interaction [12].

The reactivity of nickel for benzene hydrogenation depends strongly on the nature of the support, as the support can modify the electronic properties of the active phase and the degree of metal–support interaction can play an important role in determining catalyst activity and selectivity. For benzene hydrogenation when Ni/Nb_2O_5 was used as a catalyst, it was found to be inactive due to the very strong interaction between the metal and support [13]. Mokrane et al. [14] synthesized nickel catalysts and showed a maximum in catalyst activity with temperature. Such behavior has been observed previously and correlates to a decrease of the surface coverage by benzene at higher temperatures [15]. The activity of the catalysts was found to be strongly dependent on the metal surface area, and in particular, the small particles are found to be the most active. For example, the specific rate increased from 2.4 $mmol.min^{-1}.g_{Ni}^{-1}$ for 1.0% Ni to 20.5 $mmol.min^{-1}.g_{Ni}^{-1}$ for 0.5% Ni, while the turnover frequency (TOF) increased from 580 s^{-1} to 2,110 s^{-1}; yet, the metal surface area only increased from 26.8 $m^2.g^{-1}$ to 61.6 $m^2.g^{-1}$ indicating a particle size effect. Interestingly, over 30 years earlier, James and Moyes [16] had come to a similar conclusion from deuterium isotope studies of benzene hydrogenation and exchange over a range of metal films.

Platinum is also used as a commercial benzene hydrogenation catalyst and studies by Somorjai and coworkers have revealed that activity and selectivity can be affected by particle shape and size. Benzene hydrogenation over platinum single crystals had shown that when the reaction was performed over a Pt(1 1 1) surface, cyclohexane and cyclohexene were formed, whereas when benzene was hydrogenated over a Pt(1 0 0) surface, cyclohexane was the only product [17, 18]. This work was then followed up with a study [19] using nanocrystals of platinum with two distinct shapes: cubeoctahedral crystals with a preponderance of (1 1 1) faces, which gave cyclohexane and cyclohexene as products, and cubic nanoparticles with principally (1 0 0) faces which gave cyclohexane as the sole product, showing that the particles mirrored the single crystal surfaces. However, further testing [20] revealed an optimum particle size for activity of 2.4–3.1 nm, with larger and smaller particles being less active. Interestingly, the reaction orders did not change and were independent of particle size. These results over nickel and platinum show the complexity of benzene hydrogenation and why it has been so difficult to obtain a definitive mechanism.

4.2 Alkyl substituted aromatic hydrogenation

4.2.1 Reaction mechanisms and kinetics

It is well established that the rate of hydrogenation decreases with increasing number of methyl substituents on the aromatic ring. Toppinen and coworkers [21, 22] studied the liquid-phase hydrogenation of aromatics over nickel catalysts. From this work, it was established that the reaction rate decreased with increasing length of substituent (benzene > toluene > ethylbenzene > cumene) and also decreased with increasing number of substituents (benzene » toluene ~ xylenes > mesitylene) [23]. The position of the substituent on the aromatic ring also plays a significant role, with the *para* position being the most reactive and *ortho* position the least reactive.

The simplest alkyl substituted aromatic is toluene, with one methyl substituent on the aromatic ring, which undergoes hydrogenation to methylcyclohexane. The hydrogenation of toluene has been investigated to a much lesser extent in comparison to that of benzene; however, interest has grown significantly due to growing environmental issues regarding removal of aromatics, predominantly in the petroleum sector. At temperatures below 250°C, the reaction was found to be irreversible with negligible side reactions. The reaction was shown to go through a maximum at around 170–180°C and any further increase in temperature eventually decreased the rate [24]. This behavior was hypothesized to be due to desorption of catalytically active hydrogen from the surface of the metal at higher temperatures above 180°C, a hypothesis that was confirmed by the work of Lindfors et al. [25] via temperature-programed desorption and chemisorbed hydrogen studies.

Rahaman and Vannice [26, 27] studied the hydrogenation of toluene over palladium and found that the addition of the first two hydrogen atoms was rate determining. Also, they made an observation that hydrogen and toluene adsorb on different active metal sites. This was further confirmed by the work of Klvana et al., where they found that over nickel, toluene, and hydrogen were also adsorbed on different sites [25]. A negative reaction order was observed as the reaction rate decreased with increasing toluene partial pressure.

The addition of another methyl group to the aromatic ring brings us to the xylenes. Work has been carried out on the gas-phase catalytic hydrogenation of xylenes over various different metals including Ni, Pd, and Rh. Keane and coworkers [28, 29] studied the turnover frequencies for xylene hydrogenation at a particular temperature over a Ni catalyst and found that they decreased in the following order: *p*-xylene > *m*-xylene > *o*-xylene. The reaction order of the xylenes was all rather close to one another and ranged at around 0.1–0.44. The kinetics of the reaction over a Pd catalyst was elucidated to go in the following order: the first step involved fast adsorption of reactants followed by the rate determining surface addition of hydrogen to the aromatic. These steps give rise to a cyclic olefin intermediate, which, through adsorption–desorption, governed the overall stereochemical distribution in the final product. The reaction rate of xylenes was seen to go through a maximum in activity over a temperature range, 150–190°C, which is in agreement with previous studies.

In 2004, further research was carried out by Neyestanaki et al. [30] over a Pt-supported catalyst regarding the kinetics of xylene hydrogenation in particular *ortho*-xylene. Their work also found a zero-order dependence with respect to substrate, an order of 1.5–3 for hydrogen and a rate maximum at around 180°C. This is similar to what had been observed by other researchers studying xylene hydrogenation.

While most of the studies of alkylbenzenes have been at elevated temperatures, Jackson and Alshehri [31] studied alkylbenzene hydrogenation at low temperatures (< 70°C) over a Rh/silica catalyst. As expected, they obtained an order of reactivity of toluene > ethylbenzene > propylbenzene when hydrogenated individually. Kinetic parameters were determined and are shown in Table 4.1, revealing an increasingly negative reaction order as the alkyl chain is increased in length (the hydrogen order was ~1 in all cases). The order of reactivity observed for xylene hydrogenation was different from other literature with *para*-xylene > *ortho*-xylene > *meta*-xylene. The general activity of the xylenes was on a par with toluene but the reaction orders were similar to propylbenzene.

Table 4.1: Kinetic parameters for alkyl benzenes over Rh/silica [31].

Compound	E_a (kJ.mol^{-1})	Order in organic	k^a (min^{-1}, ×10^{-3})
Toluene	31	0.1	18.3
Ethylbenzene	46	−0.4	12.4
Propylbenzene	39	−0.8	3.3
Ortho-xylene	25	−0.7	22.3
Meta-xylene	27	−0.9	18.3
Para-xylene	42	−0.6	23.4

[a] *k*, First-order rate constant measured at 50°C, 3 barg, and ~8 mmol reactant.

Competitive hydrogenation between toluene, ethylbenzene, and propylbenzene was also studied by Alshehri et al. [31] and revealed that propylbenzene was the most strongly adsorbed species, inhibiting toluene and ethylbenzene hydrogenation. When a similar experiment was conducted using the xylenes, the reactivity of all isomers decreased. The effect of replacing hydrogen with deuterium was also studied and an inverse kinetic isotope effect (KIE) was observed for the alkylbenzenes and the xylenes except for *ortho*-xylene, which exhibited a normal KIE. This was deemed to be a secondary inverse KIE. This can be seen when there is a change in hybridization of the carbon (C–H) from sp^2 to sp^3 as would be the case in hydrogenation of the aromatic ring. With *ortho*-xylene exhibiting a positive KIE while all other alkylbenzenes exhibit a negative KIE, it raises the question of whether the rate-determining step of *ortho*-xylene is different from that of toluene and *meta*- and *para*-xylene.

During the reaction between toluene and deuterium, it was observed, using ^{2}H-NMR, that the methyl protons exchanged faster than the aromatic protons [31]. This behavior over rhodium is similar to that found over nickel, where both aliphatic and aromatic hydrogen exchanged but the rate of aliphatic exchange was over an order of magnitude faster [32]. In contrast, the exchange process with the xylenes revealed only aliphatic exchange similar to that found with palladium

[33]. The methyl groups in the xylenes showed a deuterium distribution (Figure 4.2) with *para*-xylene having the highest amount of H-6 species, which is similar to the distribution found over nickel for the exchange process [32].

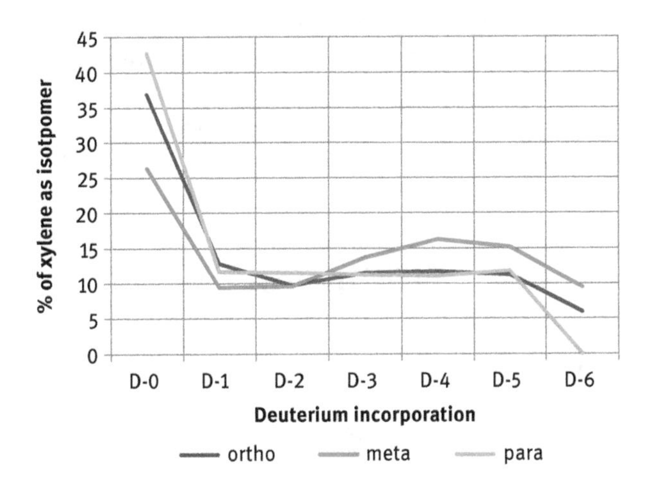

Figure 4.2: Extent of isotope exchange in each xylene after 15 min reaction [31].

4.2.2 Stereochemistry

Stereoselectivity of the dialkylcyclohexane must be considered when studying alkyl substituted aromatics as different isomers can be formed. Since 1922, it has been postulated that the hydrogenation of dialklylbenzenes would preferentially form the Z isomer in comparison to the E-isomer and indeed this is what is found experimentally. However, it is important to note that a change in temperature and hydrogen pressure can modify the Z/E ratio in the product. One of the most intriguing questions when elucidating the mechanism of dialkylbenzenes to Z/E dialkylcyclohexane is how the formation of the E-isomer takes place, since only the Z-isomer would be expected if the aromatic ring is lying flat on the surface [30]. It has been suggested that the formation of Z/E-isomers takes place via a desorption–readsorption mechanism. However in a recent study [31] of *ortho*-xylene hydrogenation, two intermediate cycloalkenes were detected 1,2-dimethylcyclohexene and 1,6-dimethylcyclohexene, which when hydrogenated would give Z-1,2-dimethylcyclohexane and E-1,2-dimethylcyclohexane, respectively. A similar explanation was proposed by Burwell et al. [34] when examining exchange processes in cycloalkanes. The effect of a competitive environment on stereochemistry was observed when xylenes were hydrogenated in each other's presence [31]. The E:Z ratio was found to increase for 1,2-dimethylcyclohexanes from ~10:1 with *ortho*-xylene hydrogenation to up to ~17:1 with *ortho*-xylene/*meta*-xylene and with all three xylenes, while the ratio for the

1,3- and 1,4-dimethylcyclohexenes did not change. This implies a subtle change in *ortho*-xylene bonding when in the presence of competing molecules that is not reflected by *meta*- or *para*-xylene.

Work carried out by Yamamoto et al. [35] studied the isomers of xylene over nickel in the vapor phase. In that system, it was found that the *ortho*- and *para*-xylene produced more than 50% E-isomer, while the *meta*-isomer produced very little E-isomer. Also, Saymeh and Asfour [36], who studied the gas-phase hydrogenation of *ortho*-xylene, reported an increase in the thermodynamically favored E-isomer when the Pd particle size was decreased.

4.3 Oxygen-substituted aromatic hydrogenation

4.3.1 Reaction mechanisms and kinetics

Phenol is the simplest oxygen-substituted aromatic and the hydrogenation pathway depends on the type of catalyst and reaction conditions. It is important to note that each metal shows unique characteristics toward product selectivity. For phenol hydrogenation, palladium is selective toward cyclohexanone, platinum is selective toward cyclohexane, while nickel, rhodium, and ruthenium are selective toward cyclohexanol. In industrial applications, phenol hydrogenation is commonly carried out over a palladium or nickel catalyst in the liquid or vapor phase, with the majority focusing their work on achieving high cyclohexanone or cyclohexanol selectivity, respectively [37]. The importance of phenol ring hydrogenation to yield cyclohexanone/cyclohexanol is related to the role of these products as intermediates in the synthesis of Nylon 6 via caprolactam and Nylon 6,6 via adipic acid. Indeed, the majority of cyclohexanone/cyclohexanol produced is used captively with only ~4% used in applications such as solvents for paints and dyes, pesticides, and as an intermediate in fine chemical and pharmaceutical manufacture. Currently, about one-third of the cyclohexanone/cyclohexanol not produced for captive conversion is produced by phenol hydrogenation; however, this amount is growing.

The hydrogenation of phenol can also be used as a pollution abatement methodology. The phenolic waste generated from a variety of different industrial sources, including oil refineries, petrochemical units, etc., is a pressing concern as phenol is an established environmental toxin. Therefore, catalytic hydrogenation of phenol and other oxygen substituted aromatics is fast emerging as an effective method to transform these hazardous substances into useful products.

The commercial significance of phenol hydrogenation to yield cyclohexanone has gathered considerable attention over the years. The two main industrial routes implemented to produce cyclohexanone involve oxidation of cyclohexane or hydrogenation of phenol [38]. The former uses homogeneous catalysts and requires high temperature and pressure, with increased undesirable by-products formed in the reaction, which can further complicate purification and lower the yield of the desired product. Whereas the latter is a gas-phase process using heterogeneous catalysts and is becoming the preferred reaction route. However, it can be difficult to achieve significant selectivity under mild conditions. The hydrogenation of phenol can either be carried out by a stepwise process where it first undergoes hydrogenation to give cyclohexanol and then subsequent dehydrogenation to give the desired product of cyclohexanone or a one-step selective process to yield cyclohexanone. The one-step method is preferred as it avoids the endothermic dehydrogenation reaction taking place and is advantageous from an efficiency standpoint. The gas-phase hydrogenation to cyclohexanone usually involves temperatures of around 150–170°C over palladium/zeolite or palladium/alumina catalysts in continuous reactors just above atmospheric pressure, while the liquid-phase hydrogenation operates at ~170°C at ~13 bar using a Pd/carbon catalyst. Both systems give yields above 90%. Direct hydrogenation to cyclohexanol in contrast is typically carried out in the vapor phase over a nickel catalyst at 120–200°C at ~20 bar. Both palladium and nickel catalysts are reported to be modified with a base to reduce the support acidity and reduce coke formation.

The proposed mechanism for phenol hydrogenation is shown in Figure 4.3.

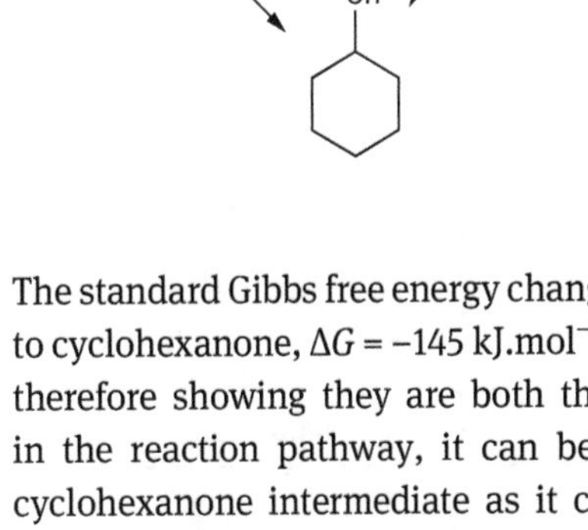

Figure 4.3: Hydrogenation of phenol.

The standard Gibbs free energy change is negative for each hydrogenation step; phenol to cyclohexanone, $\Delta G = -145$ kJ.mol^{-1}, and phenol to cyclohexanol, $\Delta G = -211$ kJ.mol^{-1}, therefore showing they are both thermodynamically favorable products. As shown in the reaction pathway, it can be complicated to achieve high selectivity to the cyclohexanone intermediate as it can be easily hydrogenated to cyclohexanol and indeed high selectivity to cyclohexanone (> 95%) still remains a challenging catalytic problem. Both, metal and support influence the selectivity of cyclohexanone [39].

The vapor-phase hydrogenation of phenol was studied by Talukdar and Bhattacharyya over palladium and platinum catalysts [40]. Their work showed that the yield of cyclohexanone generally increases with increasing temperature and also that phenol can be directly converted to cyclohexanol without going through the cyclohexanone. Tautomerism of cyclohexen-1-ol to cyclohexanone takes place readily owing to the greater stability of cyclohexanone. An increased cyclohexanol yield and corresponding decrease in the cyclohexanone/cyclohexanol ratio takes place with an increased hydrogen pressure, as there is a greater availability of hydrogen. Benzene was also seen with increased hydrogen pressure, via hydrogenolysis of phenol. On comparing the use of palladium or platinum, it was found that platinum is more active for phenol conversion, whilst palladium shows greater selectivity to cyclohexanone. Table 4.2 shows phenol conversion and cyclohexanone selectivity with both platinum and palladium catalysts. It was suggested that the behavior observed was due to strong metal–support interactions as the order of charge transfer is proposed to go in the following order Pt > Pd > Rh > Ru, which determines the extent of activation of molecules on the surface. However, it is also generally accepted that palladium has a low activity for the hydrogenation of ketones; thus, it may be expected that it could have a high activity for the hydrogenation of phenol but a low activity for the hydrogenation of cyclohexanone. For example, the activity for acetone hydrogenation was reported to go in the following order: Pt > Ni > Fe > W > Pd > Au with palladium much less active than platinum or nickel [35].

Table 4.2: Hydrogenation of phenol over platinum and palladium.

Ref.	Catalyst	Catalyst	Reaction condition				Catalytic activity (%)	
		Loading (%)	H_2/Phenol ratio	Temp. (°C)	H_2 (bar)	Time (h)	Conversion (%)	Cyclohexanone selectivity (%)
[45]	Pd/Al_2O_3	1	5.4	230	–	–	77	98
[45]	Pd/MgO	1	5.4	230	–	–	90	82
[40]	Pd/Al_2O_3	1	1:3	250	10	–	45	Product distribution: **39.5**
[46]	Pd/C LA: $AlCl_3$	5	–	50	10	7	> 99.9	> 99.9
[46]	Pd/Al_2O_3 LA:$AlCl_3$	5	–	50	10	7	> 99.9	99.3
[40]	Pt/Al_2O_3	1	1:3	250	10	–	97	Product distribution: **7.7**
[47]	Pt/CNTs	2.95	4.7	50	5	0.5	97	77.5
[48]	Pt/C	5	1:4	200	–	–	64.5	83.9

The support also has an influence; it has been proposed that when a basic MgO support is used, a nonplanar mode of phenol adsorption exists through the hydroxy group as opposed to a coplanar arrangement over an acidic Al_2O_3 support. The former should favor the addition of hydrogen in a stepwise fashion and gain high cyclohexanone selectivity [39].

The effect of adding a basic component to the catalyst formulation was studied by Scire [41]. Typically, the addition of an alkali has been to neutralize acidic sites on the supports as these can lead to carbonaceous deposits/coking and catalyst deactivation. However in this case, the addition of CaO was not purely related to minimizing deactivation by neutralizing acid sites, but also increasing the TOF by modifying the electronic character of the palladium. Such behavior has been reported for other hydrogenation reactions [42, 43]. The selectivity to cyclohexanone was deemed to be a function of achieving the correct mode of adsorption of the phenol. The proposal was that if phenol is adsorbed end-on via the hydroxyl group, then cyclohexanone is favored, but if adsorbed in a coplanar mode, then cyclohexanol is favored. In this study, the hydrogenation reaction appeared structure insensitive but, as we have seen with other reactions, a later study of phenol hydrogenation, over a series of Pd/C catalysts [44], indicated that palladium particle morphology, exposed crystallographic orientations, and electronic character all influenced selectivity and activity.

Further work on the gas-phase hydrogenation of phenol over supported nickel catalysts suggested that the reaction proceeded in a stepwise fashion with cyclohexanone forming as the partially hydrogenated product and cyclohexanol as the fully hydrogenated product. Hydrogenolysis of cyclohexanol could occur to yield cyclohexane as a product; however, only trace amounts were observed. The temperature used was found to have a considerable effect on both catalyst activity and selectivity; the conversion of phenol decreased with increasing temperature [49]. Catalyst development has been widely documented, and in 2009, dual supported palladium–Lewis acid catalysts were shown to deliver high conversion and high yield [46]. A Pd/C catalyst coupled with $AlCl_3$ (a Lewis acid) resulted in phenol conversion of > 99.9% and selectivity to cyclohexanone of > 99.9% even at temperatures as low as 50°C. Other supported palladium catalysts (Al_2O_3, NaY) with different Lewis acids ($SnCl_2$, $InCl_2$, and $ZnCl_2$) were also tested [47]. The proposed mechanism is shown in Figure 4.4, where the Lewis acid is used to stabilize the ketone and inhibit further hydrogenation. However, the proposed heterolytic cleavage of hydrogen on palladium would be unusual unless also moderated by the Lewis acid.

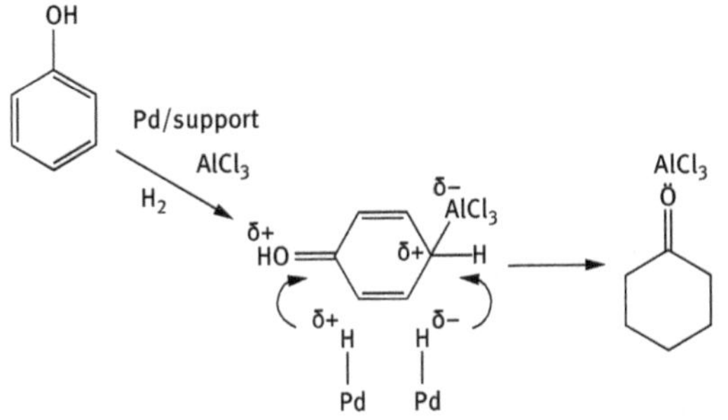

Figure 4.4: Mechanism of phenol hydrogenation using supported palladium catalysts and Lewis acids. Adapted from Ref. [47].

The kinetics of phenol hydrogenation was studied over a palladium catalyst by Mahata and Vishwanathan in the vapor phase [39]. A direct relationship was found between the rate and partial pressure of reactants; at 200°C, the order of reaction with phenol was negative and positive with hydrogen. The negative order in substrate indicated that it was strongly adsorbed on the catalyst surface; however, with an increase in temperature, the order in phenol increased indicating a weakening in the strength of adsorption. A reaction mechanism was proposed whereby the strongly adsorbed phenol reacted with dissociatively adsorbed hydrogen in a stepwise fashion to give cyclohexanone. The rate-controlling step was believed to be the surface reaction between the strongly bound phenol and weakly adsorbed hydrogen atom. The activation energy for phenol to cyclohexanone has been found by many researchers to be in the range of 30–68 $kJ.mol^{-1}$.

A study of phenol hydrogenation over a rhodium catalyst [50] revealed a propensity for hydrogenolysis. The reaction was studied in a batch reactor at low temperature < 70°C and 2–5 barg hydrogen pressure. The principal product in the early stages of reaction was cyclohexanone although cyclohexanol was produced by direct hydrogenation. However, a 20% yield of cyclohexane was also obtained (Figure 4.5) by hydrogenolysis. Kinetic analysis revealed a zero order in phenol, first order in hydrogen, and an activation energy of 27 ± 5 $kJ.mol^{-1}$. The study also examined the effect of replacing hydrogen with deuterium. An inverse KIE was found for the production of cyclohexanone suggesting that the rate-determining step for phenol hydrogenation to cyclohexanone is not hydrogen addition, whereas a positive KIE was found for the production of cyclohexane by hydrogenolysis. Hydrogenation of anisole in the same study [50] also exhibited similar behavior with an inverse KIE for hydrogenation of the aromatic ring but a positive KIE for hydrogenolysis products.

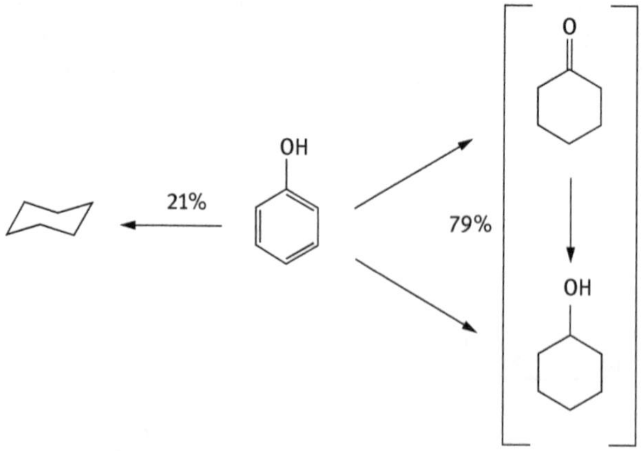

Figure 4.5: Mechanism of phenol hydrogenation and hydrodeoxygenation over Rh/silica [51].

4.4 Nitrogen-substituted aromatic hydrogenation

4.4.1 Reaction kinetics and mechanism

The simplest nitrogen substituted aromatic is aniline with a NH_2 group attached to the aromatic ring. The majority of aniline produced is used in polyurethanes manufacture but around 7,000 t of cyclohexylamine is produced by hydrogenation annually. Most of the metals from groups 9 and 10 will catalyze the hydrogenation but commercially catalysts based on nickel and cobalt are favored. A typical process would use a Co/silica catalyst, possibly modified with manganese or calcium oxides, in a slurry-phase reaction at ~230°C and ~50 bar. The main issue is selectivity as dicyclohexylamine can also be formed as a by-product.

Aniline hydrogenation has gathered some attention in the academic literature, although it should be stated that this is still rather limited. Early work on aniline hydrogenation to the subsequent ring hydrogenation product, cyclohexylamine, was carried out by Winans in 1940 [51]. The work by Winans aimed to review the choice of catalysts with a view to optimizing selectivity toward cyclohexylamine. Numerous catalysts were compared at temperatures between 200°C and 285°C including nickel on kieselguhr, Raney nickel, Raney cobalt, pure cobalt, and technical cobalt oxide. The best selectivity of > 70% to the ring hydrogenated product was seen over the cobalt oxide catalyst activated by powdered calcium oxide, which is not dramatically different from the typical industrial catalyst used today.

Further work in 1964 was carried out by Greenfield [52], where he mentioned that cobalt had been widely used with success but at high temperatures. Therefore with that in mind, Greenfield decided to focus his work on the use of noble metals as these are well known to be good at activating hydrogen under milder conditions. He examined the use of rhodium, ruthenium, platinum, and palladium for aniline hydrogenation and indeed ruthenium and rhodium were active at much lower temperatures (120–140°C) than typically used with cobalt and nickel. Platinum and palladium however were found to be rapidly poisoned by the product cyclohexylamine. Ruthenium was found to be the superior metal for the hydrogenation of aniline to cyclohexylamine.

Hydrogenation of aniline yields many different products as well as cyclohexylamine, including dicyclohexylamine, *N*-phenylcyclohexylamine, diphenylamine, ammonia, benzene, cyclohexane, cyclohexanol, and cyclohexanone (the last two formed from trace water). The mechanism of aniline hydrogenation to the cyclohexylamine is believed to proceed via a stepwise process with formation of enamine and imine intermediates. The overall product selectivity depends heavily on the catalyst choice and reaction conditions. Narayanan et al. [53] studied aniline hydrogenation over nickel and cobalt supported on alumina in the vapor phase at temperatures ~200°C. The main product of the reaction over both metals was *N*-phenylcyclohexylamine with dicyclohexylamine and cyclohexylamine also produced. In comparison when rhodium was used for vapor-phase aniline hydrogenation [54] at 200°C, the main products were cyclohexylamine and dicyclohexylamine. *N*-phenylcyclohexylamine and cyclohexane were produced in minor amounts and it was suggested that cyclohexane was formed via formation and subsequent hydrogenation of benzene. The data also revealed an antipathetic particle size effect for aniline hydrogenation, where the TOF increases with particle size. An antipathetic particle size effect was also observed by Sokolskii et al. for aniline hydrogenation over a Rh/alumina catalyst [55]. The liquid-phase batch process gave an optimum yield of cyclohexylamine of 79%, an activation energy of 50 ± 5 $kJ.mol^{-1}$ and zero-order kinetics in both aniline and hydrogen.

Aniline hydrogenation over Pt, Pd, and Rh catalysts was investigated using supercritical carbon dioxide as the solvent [56]. Very good activity and selectivity to cyclohexylamine was obtained from a 5% Rh/alumina catalyst at 80°C with a carbon dioxide pressure of 8 MPa and a hydrogen pressure of 4 MPa. Under those conditions, values of 96% for activity and 93% for selectivity were obtained. Interestingly, carbamic acid was formed from reaction between cyclohexylamine and carbon dioxide and at lower temperatures (< 60°C) the acid precipitated on the catalyst-blocking sites and lowering activity. The authors repeated the experiment at 80°C using hexane and ethanol as solvents to compare against the carbon dioxide but observed lower activity and selectivity.

The hydrogenation of alkylanilines has had only limited study. However, in a series of papers, Hindle et al. [57, 58] studied the hydrogenation of *para*-toluidine (4-methylaniline) and 4-*t*-butylaniline over rhodium/silica catalysts in the liquid phase in both stirred tank and trickle-bed reactors. They found that the reaction exhibited an antipathetic particle size effect, suggesting that the plane-face surface atoms including C^3_9 sites were active for ring hydrogenation. The main products of the hydrogenations were *cis*- and *trans*-4-methylcyclohexylamine and *cis*- and *trans*-4-*t*-butylcyclohexylamine. The proposed mechanism for this reaction was via an enamine/imine intermediate species undergoing tautomeric equilibrium with one another, which was in agreement for the mechanism proposed for aniline hydrogenation to cyclohexylamine. The authors showed that it was possible to alter the *cis*/*trans* ratio of the alkylcyclohexylamines by using solvents with different dielectric constants. It was proposed that more polar solvents, such as methanol, altered the tautomeric equilibrium between the immine and enamine intermediates (Figure 4.6) stabilizing the more polar imine species and so increasing the yield of *cis*-4-alkylcyclohexylamine over the thermodynamically more favorable *trans*-isomer. Kinetic analysis revealed zero order in alkylaniline and first order in hydrogen with activation energies of 62 ± 4 $kJ.mol^{-1}$ for the *trans*-isomer and 51 ± 6 $kJ.mol^{-1}$ for the *cis*-isomer. Poisoning of the catalyst by the products was also confirmed for these systems. The production of an aliphatic amine with a much lower pK_a resulted in strong adsorption sufficient to inhibit aromatic amine hydrogenation.

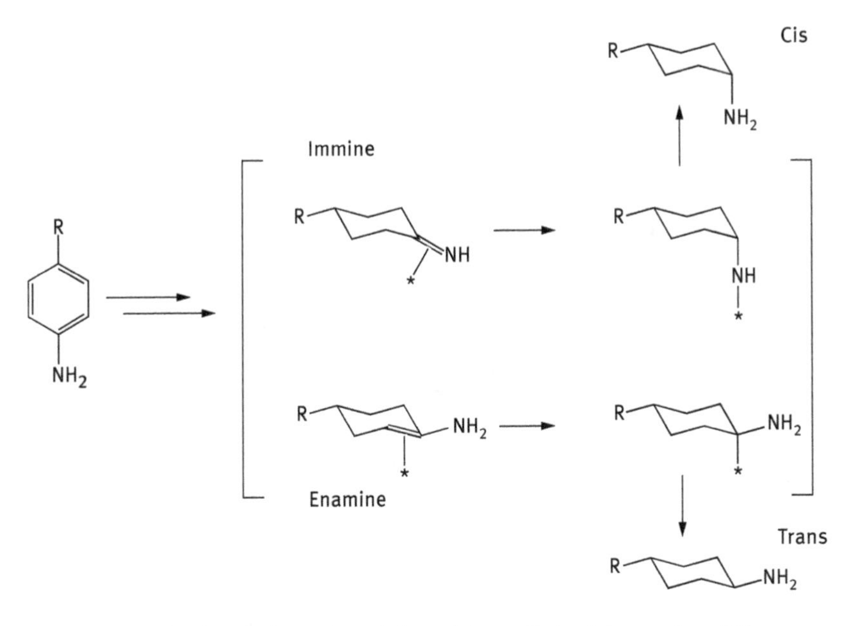

Figure 4.6: Mechanism for formation of *cis*- and *trans*-alkylcyclohexyamines [58].

4.5 Hydrogenation of fused aromatic rings

4.5.1 Introduction

In the literature, considerable effort has been devoted to understand and optimize the hydrogenation of mono-aromatic compounds; in contrast, the literature on the hydrogenation of multi-aromatic rings is rather sparse. Indeed, the number of studies that focus on hydrogenating polyaromatic compounds decreases as the number of aromatic rings increases. Hence, hydrogenation of more than two fused-ring aromatics is rarely investigated. Since naphthalene and its partially hydrogenated product tetralin (tetrahydronaphthalene) are the simplest polyaromatic species, they are usually chosen as representative reaction models for hydrogenating multi-ring compounds in the middle distillate fraction [59]. Hydrogenation of naphthalene to produce tetralin can be achieved under moderately forcing reaction conditions of 20–60 bar and 400°C using nickel sulfide or nickel molybdenum catalysts in a fixed-bed reactor [60]. Complete saturation to produce decalin (decahydronaphthalene, Figure 4.7) is much more difficult with Cooper and Donnis having reported that the hydrogenation of the first ring in naphthalene is 20–40 times faster than hydrogenating the second ring [61], due to the reversibility of the hydrogenation reaction and the resonance stabilization of the aromatic ring in tetralin [62].

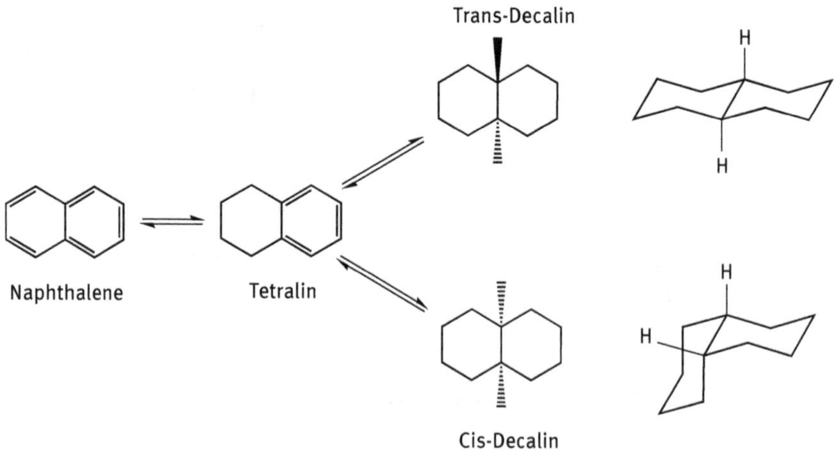

Figure 4.7: Hydrogenation of naphthalene.

4.5.2 Kinetics

Conventionally, naphthalene hydrogenation to produce decalin isomers is achieved via a deep hydro-treating process which consists of two stages. The first stage incorporates the use of sulfided CoMo, NiMo, and NiW all supported on alumina, while the second stage employs a more active hydrogenation catalyst such as platinum and palladium. The complete saturation of naphthalene incorporates tetralin as an intermediate reaction step to produce *cis*- and *trans*-decalin. The decalins are produced commercially with 97 wt% purity [63, 64]. Several studies have reported that naphthalene hydrogenation undergoes first-order reaction, while other literature has proposed a zero-order reaction, which involves octalins (octahydronaphthalenes) as important intermediates in their reaction scheme. It has been proposed that the formation of *cis*- or *trans*-decalin is solely governed by the structure of the octalin intermediates shown in Figure 4.8 [65, 66]. Hydrogenation of $\Delta^{9,10}$-octalin will yield only *cis*-decalin since the addition of hydrogen atoms can only occur on the same side. In contrast, the hydrogenation of $\Delta^{1,9}$-octalin, which should be faster, has the possibility to produce both *cis*- and *trans*-decalin depending on its orientation. *cis*-Decalin would be produced if the hydrogenation of the double bond occurred on the same side as the hydrogen atom in position 10 while *trans*-decalin will be produced if the hydrogen atom in position 10 is located in the opposite side. The double bond in the $\Delta^{1,9}$-octalin is not as stable as $\Delta^{9,10}$-octalin and hence it can be easily isomerize during the reaction to form $\Delta^{9,10}$-octalin which in turn can produce *cis*-decalin [3, 59].

In 2001 and 2002, Rautanen and coworkers did comprehensive research to study the kinetics of the liquid-phase hydrogenation of naphthalene and tetralin together with the deactivation mechanism of the catalyst used in the process [59]. The reaction

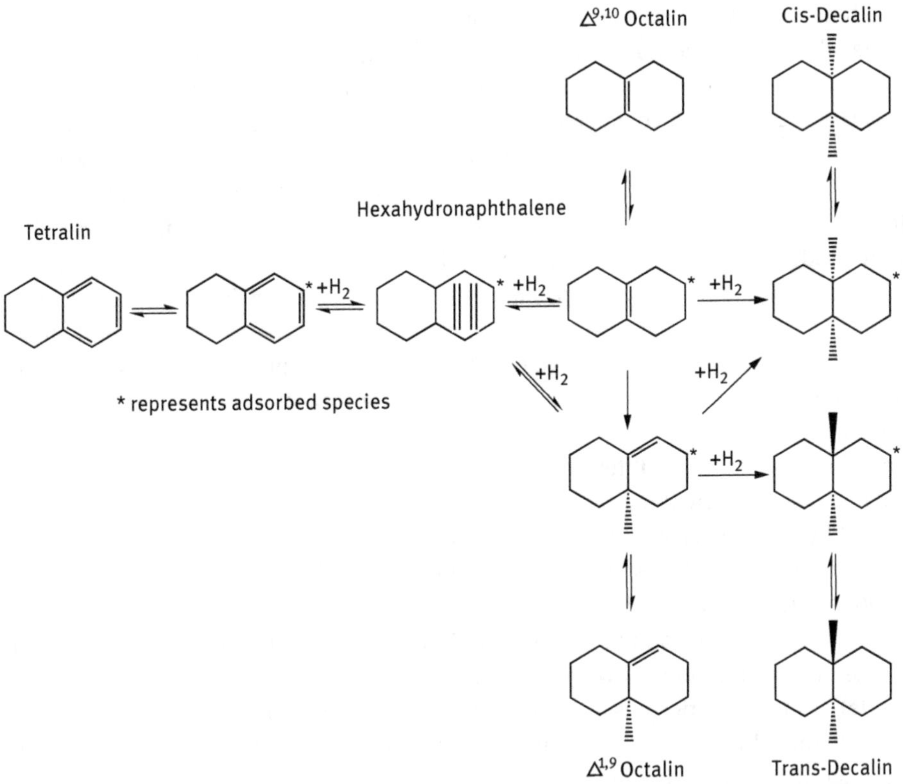

Tetralin
Hexahydronaphthalene
$\Delta^{9,10}$ Octalin
Cis-Decalin
$\Delta^{1,9}$ Octalin
Trans-Decalin
+H_2
* represents adsorbed species

Figure 4.8: Tetralin hydrogenation.

entailed hydrogenating naphthalene and tetralin in decane using a commercial 16.7 wt% Ni/γ-Al_2O_3 catalyst in three-phase Robinson Mahoney reactor (continuous stirred tank reactor, CSTR) at 353–433 K and 20–40 bar. The kinetic model proposed that the hydrogenation occurred by sequential reaction steps from naphthalene and tetralin to *cis*- and *trans*-decalins through octalins as intermediates. It assumed three adsorption modes (π-, π/σ-, and σ-adsorption) where the first two are associative (active in hydrogenation), while the third one is dissociative that leads to coke formation and catalyst deactivation. The study revealed that naphthalene hydrogenation occurs on a single active site while tetralin forms complexes on the catalyst surface with several active sites. These complexes form octalin intermediates that determine which decalin isomer will be formed (Figure 4.8). Therefore, it was proposed that catalyst deactivation decreases the hydrogenation rate of tetralin more than that of naphthalene since tetralin requires several active sites to hydrogenate. Moreover, the deactivation drives the reaction to change selectivity toward *cis*-decalin and also enhances the back reaction to naphthalene [59, 67]. Similarly, another kinetic study was performed by Romero et al. in 2008 using the same reactor under the same conditions with a difference in the reaction temperature and the presence of hydrogen sulfide (H_2S). The liquid-phase naphthalene hydrogenation was conducted over a commercial sulfided NiMo/γ-Al_2O_3 which contained 9.3 wt% Mo and 2.5 wt% Ni. The findings of this study revealed that maximum conversion (89%) was achieved at 553 K and 40 bar. The product distribution indicated that tetralin was the main hydrogenation product, whereas only small amounts of decalin were produced. Moreover, the experimental data showed the negative impact of H_2S on the hydrogenation conversion of naphthalene. A sharp drop in the conversion was observed as the H_2S partial pressures increased from 0 bar to 2 bar. The kinetic calculation was based on L–H rate equations, where two models best described the obtained experimental data. The rate-determining step in these models was associated with the third hydrogen addition from either homolytic or heterolytic hydrogen dissociation [68]. Interestingly, calculations suggested that hydrogen and sulfydril groups were the most abundant species on the surface, while the concentration of hydrocarbons was almost negligible.

Further kinetic investigation was carried out by Kirumakki and coworkers in 2006 to study hydrogenation of naphthalene over NiO/SiO_2–Al_2O_3. Several catalysts with different metal loadings were synthesized by a sol–gel method and evaluated in a fixed-bed continuous-flow quartz catalytic reactor (operated in the down-flow mode) at atmospheric pressure and 473 K. Under these conditions, the maximum conversion achieved was 88% using a catalyst with a 67 wt% Ni content; the main product was tetralin with minimal amount of decalin (Table 4.3). The kinetic model in this paper suggested that the catalyst adsorbed naphthalene twice as strongly as tetralin; nevertheless, this does not explain the reason behind the low yield of decalin, since pure tetralin was also hydrogenated and a small amount of decalin was produced. The study proposed that the low conversion of tetralin to decalin was due to the weak adsorption of tetralin on the active sites rather than the strong adsorption of naphthalene [69].

Table 4.3: Product distribution in the hydrogenation of naphthalene over NiO/SiO_2–Al_2O_3 catalysts with Si/Al ∞11 and different Ni content [70].

Ni content (wt%)	Conversion (%)	Tetralin selectivity (%)	Decalin selectivity (%)
0	0.4	0.4	0
6.8	20	20	0
11	41	40.8	0.2
15	47	46.6	0.4
19	56	55	1
40	75	71	4
67	88	76	12

4.5.3 Catalyst development

Catalyst development for this process has evolved over the past few years, where several catalysts were produced with the aim of improving the conversion and selectivity of the hydrogenation reaction under moderate conditions. Several studies were carried out to investigate the influence of different metals, supports, and preparation methodologies on the hydrogenation reaction of naphthalene and tetralin. A number of other papers examined the effects of different operating conditions and poisoned feeds on the stability and catalytic integrity of the catalysts. The alterations in the preparation technique and materials used during the catalyst synthesis were found to greatly influence catalytic activity and the outcome of the reaction. In this regard, the latest developments in this area have been reviewed and a summary of the literature tabulated in Table 4.4–4.7. The values in these tables are approximated data, which highlight the highest catalytic activities found in the literature for the hydrogenation of clean naphthalene and tetralin feeds using monometallic and bimetallic catalysts. As is often the case in catalysis, comparison between these catalysts is difficult as they were prepared by different methodologies and evaluated under different reaction conditions using alternative types of reactors but it can give some hints as to how the various parameters, such as supports, metal precursor, and reaction conditions, influence the reaction. The most commonly investigated catalysts were either noble metal catalysts, such as platinum and palladium, or nickel catalysts.

Tables 4.4–4.7 indicate that the complete hydrogenation of both naphthalene and tetralin to produce decalin can only be achieved at high hydrogen pressure using monometallic and bimetallic noble metal catalysts. The selectivity for decalin isomers is influenced by the type of noble metal and the support. Platinum showed higher selectivity toward decalin compared to palladium using the same support under the same reaction conditions. However, the type of the support can drive the reaction to favor one of decalin isomer over another. Modification of the support can alter the *trans*-to-*cis* ratio as can be observed in the case of Pt and Pd-mordenite catalysts. An increase in the silica-to-alumina ratio of the mordenite support increased the *trans*-to-*cis* ratio from 0.6:1 to 2.3:1 for platinum and 1.8:1 to 4.6:1 for palladium. Moreover, some of the supports can enhance the side reactions such as dehydrogenation and cracking as in the case of zirconium-doped mesoporous silica (Zr-MSU) support [70–72]. In terms of nickel, the investigated catalysts generally had a higher metal loading than that used with the precious metals. Nevertheless, the lowest metal loading nickel (5%) catalyst, used at a high catalyst to reactant ratio for tetralin hydrogenation, showed comparable reactivity to the noble metal catalysts under the same reaction conditions.

A study of the hydrogenation of liquid-phase naphthalene in the presence of carbon monoxide over alumina supported metals (Co, Ni, Ru, Rh, Pd, Pt) was carried out by Miura and coworkers [82]. Based on their findings, Pd showed the

Table 4.4: Naphthalene and tetralin hydrogenation over platinum catalysts.

Ref.	Catalyst	Catalyst		Reaction conditions				Catalytic activity (%)		
		Loading	Quantity	Feed	Temp. (°C)	H_2 press. (bar)	Time (h)	Conv.	*trans*-Decalin	*cis*-Decalin
[73]	**Pt/γAl_2O_3**	2%	127 mg	Naph.	280	69	1	72	3	2
[73]	**Pt/γAl_2O_3**	2%	207 mg	Naph.	200	104	2	100	44	49
[74]	**Pt/HY**	6%, 5:1 Si:Al	400 mg	Naph.	200	104	1	100	15	82
[74]	**Pt/Mord.**	6%, 17:1 Si:Al	400 mg	Naph.	200	104	1	100	37	63
[74]	**Pt/Mord.**	6%, 38:1 Si:Al	400 mg	Naph.	200	104	1	100	70	30
[75]	**Pt/AC**	2.80%	50 mg	Tetralin	100	60	0.5	89	21.2	67.8
[76]	**Pt/γ-Al_2O_3**	1%	2.5 g	Tetralin	275	35.4	6	65	49	16
[77]	**Pt/MCM-41**	2%	1.2 g	Tetralin	280	50	3	100	80	20

Table 4.5: Naphthalene and tetralin hydrogenation over palladium catalysts.

Ref.	Catalyst	Catalyst		Reaction conditions				Catalytic activity (%)		
		Loading	Quantity	Feed	Temp. (°C)	H_2 press. (bar)	Time (h)	Conv.	*trans*-Decalin	*cis*-Decalin
[73]	**Pd/γ-Al_2O_3**	2%	132 mg	Naph.	280	69	1	77	2	0.7
[73]	**Pd/γ-Al_2O_3**	2%	208 mg	Naph.	200	104	2	100	43	17
[73]	**Pd/TiO_2**	2%	144 mg	Naph.	280	69	1	84	1	0.2
[73]	**Pd/TiO_2**	2%	214 mg	Naph.	200	104	2	100	76	18
[73]	**Pd/HY**	6%, 5:1 Si:Al	400 mg	Naph.	200	104	1	100	73	27
[73]	**Pd/Mord.**	6%, 17:1 Si:Al	400 mg	Naph.	200	104	1	100	65	35
[73]	**Pd/Mord.**	6%, 38:1 Si:Al	400 mg	Naph.	200	104	1	100	82	18
[78]	**Pd/γ-Al_2O_3**	1%	10 g	Tetralin	275	35.4	6	40	32	8

Table 4.6: Naphthalene and tetralin hydrogenation over nickel catalysts.

Ref.	Catalyst	Catalyst		Reaction conditions				Catalytic activity (%)		
		Loading (%)	Quantity	Feed	Temp. (°C)	H_2 press. (bar)	Time (h)	Conv.	*trans*-Decalin	*cis*-Decalin
[69]	**$NiO/SiO_2–Al_2O_3$**	68	3 cm^3	Naph.	200	0	12	88	Decalin select. 12%	
[75]	**Ni/AC**	8.20	50 mg	Tetralin	100	60	0.5	5	0	0
[78]	**Ni/γ-Al_2O_3**	5	6.5 g	Tetralin	275	35.4	6	55	37	18
[79]	**Ni/SiO_2**	60	0.5 g	Tetralin	270	30	1.8	88.3	66	17
[70]	**Ni/Zr-MSU**	20	3 cm^3	Tetralin	275	60	0.75	100	Decalin select. 98%	

Table 4.7: Naphthalene and tetralin hydrogenation over bimetallic catalysts.

Ref.	Catalyst	Quantity catalyst	Reaction conditions				Catalytic activity (%)		
	Loading		Feed	Temp. (°C)	H_2 press. (bar)	Time (h)	Conv.	*trans*-decalin	*cis*-Decalin
[73]	**Ni–Mo/γ-Al_2O_3** 3.7% and 18.5%	207 mg	Naph.	280	69	1	17	0	0
[80]	**AuPtPd/ASA** 0.30%, 0.68%, and 0.65%	0.25 g	Naph.	210	20	5	43.7	2.2	3
[80]	**PtPd/ASA** 0.72% and 0.74%	0.25 g	Naph.	210	20	5	11.3	11.2	10
[80]	**PtPd/MWNT** 0.75% and 0.72%	0.25 g	Naph.	210	20	5	8.7	1.4	1.6
[71]	**PtPd/Zr-MSU** 1% and 1%	3 cm^3	Tetralin	315	45	7	98	72	12
[72]	**PtMo/Zr-MSU** 0.5% and 0.5%	3 cm^3	Tetralin	275	60	7	100	55.5	31.5
[72]	**PtMo/Zr-MSU** 1% and 1%	3 cm^3	Tetralin	275	60	7	100	57.8	30.3
[72]	**PtMo** (salt)**/Zr-MSU** 1% and 1%	3 cm^3	Tetralin	275	60	7	100	76.6	13.4
[75]	**PtNi/AC** 2.8% and 8.2%	50 mg	Tetralin	100	60	0.5	99	25.3	73.7
[81]	**PtPd/HDAY** 0.5% and 2%	0.2 g	Tetralin	280	40	9	100	Decalin selectivity 70%	

highest activity for naphthalene hydrogenation and no activity for carbon monoxide hydrogenation even at 473 K. In a second study [83], nickel was selected for further study. A commercial 50 wt% Ni/SiO_2–Al_2O_3 catalyst was used to hydrogenate naphthalene in a 100-ml autoclave. Naphthalene was dissolved in *n*-tridecane and the reactor was heated to temperatures in the range 303–473 K before hydrogen or 2% CO/H_2 was introduced to reactor. The results indicated that the rate of naphthalene hydrogenation increased proportionally with temperature under both pure hydrogen and 2% CO/H_2 mixture but that the temperature required to hydrogenate naphthalene in the presence of carbon monoxide was much higher than that required in hydrogen to achieve the same rate of hydrogenation. For example, the rate of naphthalene hydrogenation in the absence of carbon monoxide at 313 K was 32.2 mmol.h^{-1}.$gcat^{-1}$ while it required 418 K to achieve 33 mmol.h^{-1}.$gcat^{-1}$ in the presence of carbon monoxide. This was deemed to be due to competition for sites between the carbon monoxide and naphthalene. The main product of naphthalene hydrogenation was tetralin whereas carbon monoxide hydrogenation produced methane but at a much slower rate than naphthalene hydrogenation.

The hydrogenation of naphthalene using bimetallic catalyst systems modified by gold was studied in 2006 by Pawelec et al. [80] As mentioned earlier, monometallic noble catalysts showed high activity toward naphthalene hydrogenation in earlier studies. The high-activity-associated Pt- and Pd-based catalysts have been associated with their ability to dissociate hydrogen at lower temperatures compared to the conventional nickel sulfide or nickel–molybdenum catalysts. The study entailed the use of bimetallic catalysts that were prepared by conventional wet impregnation by loading platinum and palladium onto amorphous silica–alumina (ASA) and multiwall carbon nanotube supports, while the gold/platinum/palladium bimetallic was prepared by simultaneous reduction of metal precursors by ethanol in the presence of poly(*N*-vinyl-2-pyrrolidone). The catalysts were tested to evaluate their activities to hydrogenate naphthalene in the presence and absence of sulfur compounds. Naphthalene hydrogenation was carried out in a continuous-down-flow fixed-bed reactor at 20 bar and temperatures of 448–483 K. The highest naphthalene conversion and lowest deactivation was observed with the ternary AuPtPd/ASA catalyst Tetralin was the main product with selectivities over 90% for the AuPtPd/ASA catalyst. The authors concluded that the enhanced activity and sulfur resistance observed with the AuPtPd/ASA catalyst was related to its larger metal surface area and the “ensemble” effect of the $Au_{70}Pd_{30}$ and $Au_{42}Pd_{58}$ alloy particles formed on ASA. They proposed that presence of gold on the surface resulted in a weakening of the strength of adsorption of aromatic compounds (e.g., naphthalene/dibenzothiophene (DBT)), resulting in more facile product desorption [80].

The hydrogenation of naphthalene goes through tetralin as an intermediate, which is much more difficult to hydrogenate. Weitkamp [62] reports an order in selectivity of Pd > Pt > Rh > Ir > Ru for the hydrogenation to tetralin and notes that under the conditions used, Pd was unique in not producing decalin. Subsequent studies have shown that palladium will produce decalin from tetralin but only poorly. Density Functional theory (DFT) calculations of tetralin adsorption on Ir(1 0 0), Pt(1 0 0), and Pd(1 0 0) [84] indicated that tetralin bonding was only slightly weaker than benzene by ~15 $kJ.mol^{-1}$ for platinum and iridium but was 85 $kJ.mol^{-1}$ weaker than benzene over palladium. Indeed, the values for tetralin adsorption matched well with butylbenzene suggesting that tetralin may be better represented as a substituted benzene. The difference in hydrogenation behavior can be seen in a comparative study of the hydrogenation of tetralin over supported Ni, Pt, and Pd catalysts [78]. The catalysts were prepared using conventional incipient wetness impregnation technique to load the metals on commercial γ-Al_2O_3 with 1.0 wt% loading for Pt and Pd, and 5.0 wt% for Ni. The hydrogenation of tetralin in the vapor phase was investigated in a continuous fixed-bed reactor to measure the catalytic activities of the catalysts where the reaction conditions were set at 35.4 bar hydrogen and 548 K. The highest conversion was achieved using Pt/γ-Al_2O_3 (~65%), while the conversion over the Ni catalyst was 10% lower and the Pd catalyst 20% lower. In terms of *trans*/*cis*-decalin selectivity, the ratio continuously increased with tetralin conversion over Pt and Ni catalysts, while it remained almost constant over Pd. This change in selectivity is due to the high isomerization activity of Pt and Ni relative to Pd, which was confirmed by a decalin isomerization reaction over the three metals.

A Pt/Ni bimetallic catalyst supported on activated carbon was tested by Huang et al. [75] for the hydrogenation of solvent-free tetralin and showed a distinct advantage over the monometallic species. The catalyst was synthesized by a galvanic replacement reaction to produce a catalyst with metal loadings of 8.2 wt% Ni and 2.8 wt% Pt. The catalysts were evaluated in a stainless steel autoclave reactor under 60 bar hydrogen pressure at 373 K. The hydrogenation reaction gave 99.1% tetralin conversion with 98.6% decalin selectivity in 0.5 h. The product distribution of the hydrogenation reaction favored *cis*-decalin selectivity over *trans*-decalin with selectivities of 73.3% and 25.2%, respectively. In comparison, the monometallic Pt/AC only attained a tetralin conversion of 88.9% with 93.5% decalin selectivity, while Ni/AC achieved less than 5% conversion. However, the main benefit was in the ability to recycle the bimetallic catalyst with no loss in activity. During six consecutive cycles, the bimetallic Pt–Ni/AC lost < 5% of its activity, while the monometallic Pt/AC showed a rapid decrease in activity with a ~50% decrease in conversion after the six runs. Post-reaction analysis revealed that the Pt/AC had sintered with the platinum particle size increasing from 18 nm to 28 nm, whereas the Pt–Ni/AC catalyst did not exhibit any sintering. The reason proposed for this better stability of the bimetallic catalyst was a strong interaction of the electron-deficient Pt with both Ni and the carbon support, which would prevent leaching and agglomeration of the platinum.

4.6 Concluding remarks

From the work reviewed above, we may make some general comments concerning the hydrogenation of the aromatic ring. Although benzene hydrogenation has been used as a test reaction on numerous occasions to differentiate between catalysts, it is clear now that this is a flawed strategy. The sensitivity of benzene hydrogenation to the structure (and size) of the metal crystallite has been unambiguously demonstrated. Indeed, the evidence suggests that aromatic ring hydrogenation in general is structure sensitive with phenol, aniline, and alkylanilines all showing an antipathetic particle size dependence. Therefore, one must treat with caution studies using benzene hydrogenation as a test reaction for a range of catalysts.

There is also the question of the mechanism. Isotope studies with benzene have revealed the absence of a KIE [85]. More generally, replacing hydrogen with deuterium mostly reveals an inverse KIE for ring hydrogenation. These studies confirm that hydrogen is unlikely to be involved directly in the rate-determining step. This is a conclusion that has been known for years but rarely impacts on mechanistic discussions, which are often more concerned with the mode of ring adsorption. Indeed, a more likely rate-determining step is the conversion of the sp^2-hybridized carbon into an sp^3-hybridized carbon. This would be consistent with loss of aromatic character being the slow step. Nevertheless, it is clear with substituted aromatic species that the substituent group plays a large role in adsorption. It has long been recognized in phenol adsorption that the hydroxyl group has a significant influence but it is now clear that even methyl groups in toluene and the xylenes interact with the surface.

What is also becoming clear is that in a competitive situation, the hydrogenation behavior, activity/selectivity, of the individual reactants is not a good proxy for the competitive reaction. In benzene hydrogenation, where there is a single reactant feed, this may not be an issue but in mixed hydrogenation streams such as Py-gas or pyrolysis oil from biomass, where there are multiple aromatic species, simple modeling using a single component has no value in trying to understand the complexity of the system.

The hydrogenation of naphthalene was investigated in great detail by Weitkamp [4] and in some ways our understanding has progressed only marginally from those studies. Improved catalysts for naphthalene hydrogenation have been developed and the mechanism proposed by Weitkamp has been robust enough to be used for 50 years. However, although there are a number of DFT studies on the adsorption of naphthalene, there are few practical studies to complement this work [86, 87]. As may be expected, the adsorption energetics for naphthalene suggests flat adsorption on the surface with minor structural relaxation. The adsorption of tetralin on metals is similarly sparse. Nevertheless, there are more studies on adsorption than there are on particle size effects on the catalytic hydrogenation of naphthalene or tetralin, which are noticeably absent from the literature.

The hydrogenation of the aromatic ring is a fascinating area, which has yet to give up all its secrets.

References

[1] Sabatier, P. and Senderens, J-B. Hydrogénations directes réalisées en présence de nickel réduit : préparation de l'hexahydrobenzéne. Compt. Rend. Chimie, 1901, 132, 210; Sabatier, P, How I Have Been Led to the Direct Hydrogenation Method by Metallic Catalysts. Industrial and Engineering Chemistry. 1926, 18(10), 1005–1008.

[2] Stanislaus, A, Cooper, BH. Aromatic Hydrogenation Catalysis: A Review. Catalysis Reviews-Science and Engineering. 1994, 36, 75–126.

[3] Bond, G. Metal Catalysed reactions of Hydrocarbons, Springer, New York, 2005.

[4] Kistiakowsky, G, Ruhoff, JR., Smith, HA., Vaughan, WE. Heats of Organic Reactions III. Hydrogenation of Some Higher Olefins. Journal of the American Chemical Society. 1936, 58(1), 137–45.

[5] Kistiakowsky, G, Ruhoff, JR, Smith, HA, Vaughan, WE, Heats of Organic Reactions. IV. Hydrogenation of Some Dienes and of Benzene. Journal of American Chemical Society, 1936, 58(1), 146–53.

[6] Aramendía, M, Borau, V, Jiménez, C, Marinas, JM., Rodero, F, Sempere, ME. Hydrogenation of Xylenes Over Supported Pt Catalysts. Influence of Different Variables on their Catalytic Activity. Reaction Kinetics and Catalysis Letters. 1992, 46(2), 305–12.

[7] Bancroft, WD., George, AB. Hydrogenation of Benzene with Nickel and Platinum. The Journal of Physical Chemistry. 1931, 35(8), 2219–25.

[8] Taylor, W, Staffin, HK. Kinetics of the Hydrogenation of Benzene Over Supported Cobalt. The Journal of Physical Chemistry. 1967, 71(10), 3314–9.

[9] Keane, M, Patterson, PM. Compensation Behaviour in the Hydrogenation of Benzene, Toluene and o-xylene over Ni/SiO_2. Determination of True Activation Energies. Journal of the Chemical Society, Faraday Transactions. 1996, 92(8), 1413–21.

[10] Mirodatos, C, Dalmon, JA, Martin, GA. Steady-State and Isotopic Transient Kinetics of Benzene Hydrogenation on Ni Catalysts. Journal of Catalysis. 1987, 105, 405–15.

[11] Smeds, S, Salmi, T, Murzin, D. Gas Phase Hydrogenation of o- and p-xylene on Ni/Al_2O_3. Kinetic Modelling. Applied Catalysis A. General. 1997, 150(115), 115–29.

[12] Savva, P, Goundani, K, Vakros, J, Bourikas, K, Fountzoula, C, Vattis, D, Lycourghiotis, A, Kordulis, C. Benzene Hydrogenation over Ni/Al_2O_3 Catalysts Prepared by Conventional and sol–gel Techniques. Applied Catalysis B: Environmental. 2008, 79(3), 199–207.

[13] Jasik, A, Wojcieszak, R, Monteverdi, S, Ziolek, M, Bettahar, MM., Study of Nickel Catalysts Supported on Al_2O_3, SiO_2 or Nb_2O_5 Oxides. Journal of Molecular Catalysis A: Chemical. 2005, 242(1–2), 81–90.

[14] Mokrane, T, Boudjahem, A-G, Bettahar, M. Benzene Hydrogenation over Alumina-Supported Nickel Nanoparticles Prepared by Polyol Method. RSC Advances. 2016. 6(64). 59858–64.

[15] Chettibi, M, Boudjahem, A, Bettahar, M. Synthesis of Ni/SiO_2 Nanoparticles for Catalytic Benzene Hydrogenation. Transition Metal Chemistry. 2011, 36(2), 163–9.

[16] James, R, Moyes, RB. Patterns of Activity in the Benzene–Deuterium Exchange Reaction and the Hydrogenation of Benzene Catalysed by Evaporated Metal Films. Journal of the Chemical Society, Faraday Transactions. 1978, 74(0), 1666–75.

[17] Bratlie, K, Flores, LD., Somorjai, GA. In Situ Sum Frequency Generation Vibrational Spectroscopy Observation of a Reactive Surface Intermediate during High-Pressure Benzene Hydrogenation. Journal of Physical Chemistry. 2006, 110(20), 10051–7.

[18] Bratlie, K, Kliewer, CJ, Somorjai, GA. Structure Effects of Benzene Hydrogenation Studied with Sum Frequency Generation Vibrational Spectroscopy and Kinetics on Pt(111) and Pt(100) Single-Crystal Surfaces. Journal of Physical Chemistry. 2006, 110(20), 17925–30.

[19] Bratlie, K, Lee, H, Komvopoulos, K, Yang, P, Somorjai, GA. Platinum Nanoparticle Shape Effects on Benzene Hydrogenation Selectivity. Nano Letters. 2007, 7(10), 3097–101.
[20] Pushkarev, V, An, K, Alayoglu, S, Beaumont, SK., Somorjai, GA. Reforming of C 6 Hydrocarbons Over Model Pt Nanoparticle Catalysts. Journal of Catalysis. 2012, 55(11–13), 723–30.
[21] Toppinen, S, Rantakyla T-K, Salmi, T, Aittamaa, J. Kinetics of the Liquid-Phase Hydrogenation of Disubstituted and Trisubstituted Alkylbenzenes over a Nickel-Catalyst. Industrial & Engineering Chemistry Research. 1996, 35 (12), 4424–33.
[22] Toppinen, S, Rantakyla T-K, Salmi, T, Aittamaa, J. Liquid-Phase Hydrogenation Kinetics of Aromatic Hydrocarbon Mixtures. Industrial & Engineering Chemistry Research, 1997, 36(6), 2101–9.
[23] Murzin, D, Smeds, S, Salmi, T. Kinetics and Stereoselectivity in Gas-Phase Hydrogenation of Alkylbenzenes Over Ni/Al_2O_3. Reaction Kinetics and Catalysis Letters. 2000, 71(1), 47–54.
[24] Karanth, NG., Hughes, R. The Kinetics of the Catalytic Hydrogenation of Toluene. Journal of Chemical Technology and Biotechnology. 1973, 23(11), 817–27.
[25] Lindfors, L, Salmi, T, Smeds, S. Kinetics of Toluene Hydrogenation on Ni/Al_2O_3 Catalyst. Chemical Engineering Science. 1993, 48(22), 3813–28.
[26] Bahaman, MV., Vannice, MA. The Hydrogenation of Toluene and o-, m-, and p-Xylene over Palladium: I. Kinetic Behavior and o-xylene Isomerization. Journal of Catalysis. 1991, 127(1), 251–66.
[27] Rahaman, MV., Vannice, MA. The Hydrogenation of Toluene and o-, m-, and p-xylene over Palladium II. Reaction Model. Journal of Catalysis. 1991, 127(1), 267–75.
[28] Keane, M, Patterson, PM. Compensation Behaviour in the Hydrogenation of Benzene, Toluene and o-Xylene over Ni/SiO_2. Determination of True Activation Energies. Journal of American Chemical Society. 1996, 92(8), 1413–21.
[29] Keane, M, Patterson, PM. The Role of Hydrogen Partial Pressure in the Gas-Phase Hydrogenation of Aromatics over Supported Nickel. Industrial & Engineering Chemistry Research. 1999, 38(4), 1295–305.
[30] Neyestanaki, A, Backman, K, Mäki-Arvela, P. Wärna, J, Salmi, D, Murzin, D.Yu. Kinetics and Modeling of o-xylene Hydrogenation over $Pt/\gamma\text{-}Al_2O_3$ Catalyst. Chemical Engineering Journal. 2003, 91(2–3), 271–8.
[31] Alshehri, F, Weinart, HM., Jackson, SD. Hydrogenation of Alkylaromatics over Rh/silica. Reaction Kinetics, Mechanisms and Catalysis. 2017, 122(2), 699–714.
[32] Crawford, E, Kemball, C. Catalytic Exchange of Alkylbenzenes with Deuterium on Nickel Films. Transactions of the Faraday Society. 1962, 58, 2452–67.
[33] Williams, P, Than, C, Rabbani S, Long, MA., Garnett, JL. Heterogeneous Palladium-Catalyzed Exchange Labelling of Representative Organic Compounds with tritium gas. Journal of Labelled Compounds and Radiopharmaceuticals. 1995, 36(1), 1–14.
[34] Burwell, R, Jr. Shim, BKS., Rowlinson, HC. The Exchange between Hydrocarbons and Deuterium on Palladium Catalysts. Journal of the American Chemical Society. 1957, 79(19), 5142–8.
[35] Yamamoto, H, Kwan, T. Selectivity of Group VIII Metals for the Catalytic Hydrogenation and Hydrogenolysis of Phenol. Chemical & Pharmaceutical Bulletin. 1969, 17(6), 1081–9.
[36] Saymeh, RA., Asfour, HM. Kinetic Study of the Gas Phase Hydrogenation of O-Xylene Over Pt/Al_2O_3 Catalyst. Oriental Journal of Chemistry. 2000, 16(1).
[37] Donato, A, Neri, G, Pietropaolo, R. Hydrogenation of Phenol to Cyclohexanone over Pd/MgO. Journal of Chemical Technology and Biotechnology. 1991, 51(2), 145–53.
[38] Wang, Y, Yao, J, Li, H, Su, D, Antonietti, M. Highly Selective Hydrogenation of Phenol and Derivatives over a Pd@Carbon Nitride Catalyst in Aqueous Media. Journal of the American Chemical Society. 2011, 133(8), 2362–5.
[39] Mahata, N, Vishwanathan, V. Kinetics of Phenol Hydrogenation over Supported Palladium Catalyst. Journal of Molecular Catalysis A: Chemical. 1997, 120(1), 267–70.

[40] Talukdar, A, Bhattacharyya, KG. Hydrogenation of Phenol over Supported Platinum and Palladium Catalysts. Applied Catalysis A: General. 1993, 96(2), 229–39.
[41] Scirè, S, Minicò, C, Crisafulli, C. Selective Hydrogenation of Phenol to Cyclohexanone over Supported Pd and Pd-Ca Catalysts: An Investigation on the Influence of Different Supports and Pd Precursors. Applied Catalysis A: General. 2002, 235(21), 21–31.
[42] Gusovius, A, Walting, TC, Prins, R. Ca Promoted Pd/SiO2 Catalysts for the Synthesis of Methanol from CO: The Location of the Promoter. Applied Catalysis A: General. 1999, 188(1–2), 187–99.
[43] Pillai, U, Sahle-Demessie, E. Strontium as an Efficient Promoter for Supported Palladium Hydrogenation Catalysts. Applied Catalysis A: General. 2005, 281(1–2), 31–8.
[44] Park, C, Keane, MA. Catalyst Support Effects: Gas-phase Hydrogenation of Phenol Over Palladium. Journal of Colloid and Interface Science. 2003, 266(1), 183–94.
[45] Zhong, J, Chen, J, Chen, L. Selective Hydrogenation of Phenol and Related Derivatives. Catalysis Science and Technology. 2014, 4(10), 3555–69.
[46] Lui, H, Jiang, T, Han, B, Liang, S, Zhou, Y. Selective Phenol Hydrogenation to Cyclohexanone over a Dual Supported Pd-Lewis Acid Catalyst. Science. 2009, 326(5957), 1250–2.
[47] Li, F, Cao, B, Zhu, W, Song, H, Wang, K, Li, C. Hydrogenation of Phenol over Pt/CNTs: The Effects of Pt Loading and Reaction Solvents. Catalysts. 2017, 7(5).
[48] Srinivas, S, Rao, PK. Highly selective Pt–Cr/C alloy Catalysts for Single-Step Vapour Phase Hydrogenation of Phenol to Give Cyclohexanone. Journal of the Chemical Society, Chemical Communications. 1993, 0(1), 33–4.
[49] Shin, E, Keane, MA. Gas-Phase Hydrogenation/Hydrogenolysis of Phenol over Supported Nickel Catalysts. Industrial & Engineering Chemistry Research. 2000, 39(4), 883–92.
[50] Alshehri, F. The Hydrogenation of Substituted Benzenes over Rh/silica. 2017. University of Glasgow, PhD thesis.
[51] Winans, C. Hydrogenation of Aniline. Industrial & Engineering Chemistry. 1940, 32(9), 1215–6.
[52] Greenfield, H. Hydrogenation of Aniline to Cyclohexylamine with Platinum Metal Catalysts. Journal of Organic Chemistry. 1964, 29(10), 3082–4.
[53] Narayanan, S, Unnikrishnan, RP. Comparison of Hydrogen Adsorption and Aniline Hydrogenation over co-Precipitated Co/Al_2O_3 and Ni/Al_2O_3 Catalysts. Journal of the Chemical Society, Faraday Transaction. 1997, 93(10), 2009–13.
[54] Vishwanathan, V, Sajjad, SM., Narayanan, S. Gas Phase Aniline Hydrogenation over Supported Rhodium/Alumina Catalyst. Indian Journal of Chemistry. 1991, 30A, 679–81.
[55] Sokolskii, D, Ualikhanova, A, Temirbulatova, AE. Aniline Hydrogenation in the Presence of Alumina-Supported Rhodium under Hydrogen Pressure. Reaction Kinetics and Catalysis Letters. 1982, 20(1–2), 35–7.
[56] Chatterjee, M, Sato, M, Kawanami, H, Ishizaka, T, Yokoyama, T, Suzuki, T. Hydrogenation of Aniline to Cyclohexylamine in Supercritical Carbon Dioxide: Significance of Phase Behaviour. Applied Catalysis A: General. 2011, 396(1–2), 186–93.
[57] Hindle, KT., Jackson, SD., Stirling, D, Webb, G. The Hydrogenation of Para-toluidine over Rhodium/Silica: The Effect of Metal Particle Size and Support Texture. Journal of Catalysis. 2006, 241(2), 417–25.
[58] Graham, KF., Hindle, KT., Jackson, SD., Williams, DJM., Wuttke, S. Stereoselective Synthesis of Alicylic Amines. Topics in Catalysis. 2010, 53(15), 1121–5.
[59] Rautanen, PA., Aittamaa, JR., Krause, AOI. Liquid phase Hydrogenation of Tetralin on Ni/Al_2O_3. Chemical Engineering Science. 2001, 56(4), 1247–54.
[60] Rase, HF. Handbook of Commercial Catalysts: Heterogeneous Catalysts. 2000.
[61] Cooper, B, Donnis, B. Aromatic Saturation of Distillates: An Overview. Applied Catalysis A: General. 1996, 137(2), 203–23.

[62] Weitkamp, A. Deuteriation and Deuterogenation of Naphthalene and Two Octalins. Journal of Catalysis. 1966, 6(3), 431–57.
[63] Mason, RT. Naphthalene, Kirk-Othmer Encyclopedia of Chemical Technology. 2002.
[64] Technical Resources: Tetralin and Decalin information. 1992.
[65] Sapre, A, Gates, B. Hydrogenation of Aromatic Hydrocarbons Catalyzed by Sulfided Cobalt Oxide-Molybdenum Oxide. Alpha.-Aluminum Oxide. Reactivities and Reaction Networks. Industrial and Engineering Chemistry Process Design and Development. 1981, 20(1), 68–73.
[66] Huang, T, Kang, B., Kinetic Study of Naphthalene Hydrogenation over Pt/Al_2O_3 Catalyst. Industrial and Engineering Chemistry Research. 1995, 34(4), 1440–8.
[67] Rautanen, P, Lylykangas, M, Aittamaa, J, Krause, A. Liquid-Phase Hydrogenation of Naphthalene and Tetralin on Ni- Al_2O_3 Kinetic Modeling. Industrial & Engineering Chemistry Research. 2002, 41(24), 5966–75.
[68] Romero, C, Thybaut, J, Marin, G. Naphthalene Hydrogenation over a NiMo Al_2O_3 Catalyst. Experimental Study and Kinetic Modelling. Catalysis Today. 2008, 130(1), 231–42.
[69] Kirumakki, S, Shpeizer, B, Sagar, G, Chary, K, Clearfield, A. Hydrogenation of Naphthalene over $NiO\text{-}SiO_2\text{–}Al_2O_3$ catalysts. Structure–Activity Correlation. Journal of Catalysis. 2006, 242(2), 319–31.
[70] Eliche-Quesada, D, Merida-Robles, J, Maireles-Torres, P, Rodrıguez-Castellon, E, Jimenez-Lopez, A. Supported Nickel Zirconium-Doped Mesoporous Silica Catalysts. Influence of the Nickel Precursor. Langmuir. 2003, 19(12), 4985–91.
[71] Carrion, M, Manzano, BR., Jalon, FA., Eliche-Quesada, D, Maireles-Torres, P, Rodrıguez-Castellon, E, Jimenez-Lopez, A. Influence of the Metallic Precursor in the Hydrogenation of Tetralin over Pd–Pt Supported Zirconium Doped Mesoporous Silica. Green Chemistry. 2005, 7(11), 793–9.
[72] Carrion, M, Manzano, BR., Jalon, FA., Maireles-Torres, P, Rodrıguez-Castellon, E, Jimenez-Lopez, A. Hydrogenation of Tetralin over Mixed PtMo Supported on Zirconium Doped Mesoporous Silica. Use of Polynuclear Organometallic Precursors. Journal of Molecular Catalysis A: Chemical. 2006, 252(9), 31–9.
[73] Lin, S, Song, C. Noble Metal Catalysts for Low-Temperature Naphthalene Hydrogenation in the Presence of Benzothiophene. Catalysis Today. 1996, 31 (1–2), 93–104.
[74] Schmitz, A, Song, C. Shape-Selective Isopropylation of Naphthalene. Reactivity of 2, 6-Diisopropylnaphthalene on Dealuminated Mordenites. Catalysis Today. 1996, 31(1–2), 19–25.
[75] Huang, Y, Ma, Y, Cheng, Y, Wang, L, Li, X. Supported Nanometric Platinum–Nickel Catalysts for Solvent-Free Hydrogenation of Tetralin. Catalysis Communications. 2015, 69, 55–8.
[76] Dokjampa, S, Rirksomboon, T, Osuwan, S, Jongpatiwut, S, Resasco, S. Comparative Study of the Hydrogenation of Tetralin on Supported Ni, Pt, and Pd Catalysts. Catalysis Today. 2007, 123 (1–4), 218–23.
[77] Luo, M, Wang, Q, Li, G, Zhang, X, Wang, L, Jiang, T. Enhancing Tetralin Hydrogenation Activity and Sulphur-Tolerance of Pt MCM-41 Catalyst with $Al(NO_3)_3$, $AlCl_3$ and $Al(CH_3)_3$. Catalysis Science & Technology. 2014, 4, 2081–90.
[78] Dokjampa, S., Rirksomboon, T, Osuwan, S, Jongpatiwut, S, Resasco, D. Comparative Study of the Hydrogenation of Tetralin on Supported Ni, Pt, and Pd Catalysts. Catalysis Today. 2007, 123(1–4), 218–23.
[79] Upare, D, Song, B, Lee, C. Hydrogenation of Tetralin over Supported Ni and Ir Catalysts. Journal of Nanomaterials. 2013, Article ID 210894.
[80] Pawelec, B, Parola, VL., Thomas, S, Fierro, JLG. Enhancement of Naphthalene Hydrogenation over $PtPd\text{-}SiO_2\text{-}Al_2O_3$ Catalyst Modified by Gold. Journal of Molecular Catalysis A: Chemical. 2006, 253(1–2), 30–43.

[81] Liu, H, Meng, X, Zhao, D, Li, Y. The Effect of Sulfur Compound on the Hydrogenation of Tetralin over a Pd–Pt HDAY Catalyst. Chemical Engineering Journal. 2008, 140, 424–31.
[82] Suzuki, T, Sekine, H, Ohshima, M, Kurokawa, H, Miura, H. Hydrogenation of Naphthalene and Tetralin in the Presence of CO over Various Supported Metal Catalysts. Kagaku Kougaku Ronbunshu. 2007, 33(6), 593–8.
[83] Sekine, H, Ohshima, M, Kurokawa, H, Miura, M. Liquid Phase Hydrogenation of Naphthalene in the Presence of CO Over Supported Ni Catalyst. Reaction Kinetics and Catalysis Letters. 2008, 95(1), 99–105.
[84] Li, X, Wong, MSM, Lim, KH. Adsorption of Tetralin and Hydrogenated Intermediates and Products on the (100) Surfaces of Ir, Pt and Pd: A DFT Study. Theoretical Chemistry Accounts, 2010, 127, 401–9.
[85] Meerten, R., Morales, A, Barbier, J, Maurel, R. Isotope Effects in the Hydrogenation and Exchange of Benzene on Platinum and Nickel. Journal of Catalysis. 1979, 58(1), 43–51.
[86] Gottfried, J., Vestergaard, EK., Bera, P, Campbell, CT. Heat of Adsorption of Naphthalene on Pt (111) Measured by Adsorption Calorimetry. Journal of Physical Chemistry B. 2006, 110(35) 17539–45.
[87] Jenkins, S. Aromatic Adsorption on Metals via First-Principles Density Functional Theory. Proceedings of the Royal Society. 2009.

Alan M. Allgeier and Sourav K. Sengupta

5 Nitrile hydrogenation

5.1 Introduction

The metal catalyzed hydrogenation of nitriles to amines represents a synthetically and commercially valuable transformation [1]. As with most hydrogenation reactions, it is highly atom-efficient and applicable across a range of scales from pharmaceuticals to surfactants to Nylon 6,6 manufacture. Catalytic nitrile hydrogenation has been known since at least the time of Sabatier [2], who described utilization of reduced nickel catalysts, a class of materials, which still dominate in practical applications [3]. This chapter provides a perspective on the industrial significance of nitrile hydrogenation and reviews the current level of mechanistic understanding, the influence of reaction parameters, and the types of catalysts, both heterogeneous and homogeneous, useful for nitrile hydrogenation. Other thorough reviews have been published covering aspects of liquid-phase hydrogenation [3], mechanism of nitrile reduction on solid catalyst surfaces [4], and recent developments in soluble transition metal catalysts [5]. Accordingly, this treatment seeks not to be exhaustive but rather to describe the state of the field, with particular focus on industrial relevance. The chapter closes with an outlook on current trends and future needs for research on nitrile hydrogenation.

5.2 Applications and significance of nitrile hydrogenation

Chemoselectivity is perhaps one of the most important criteria in the catalytic hydrogenation of nitriles. This class of reactions is complicated due to the multitude of reaction pathways that lead to the formation of primary, secondary, and tertiary amines. Therefore, depending on which one of these amines is the desired final product of the process, serious consideration must be given to the selection and/or design of the catalysts. Additionally, the choice of solvent, catalyst modifiers, operating conditions, and the type of reactor are also important factors. All of these must be necessarily looked into, to make the process robust and commercially viable, to provide an acceptable rate, as well as to minimize the formation of by-products so that the catalyst can sustain itself and survive over a long period of time and/or be recycled and reused. This section is not meant to be an exhaustive review of nitrile hydrogenation applications, but an overview of the interplay of some of the above factors that make the study of catalytic hydrogenation of nitriles interesting and industrially significant.

https://doi.org/10.1515/9783110545210-005

5.2.1 1,6-Hexamethylenediamine synthesis

When it comes to nitrile hydrogenation, by far the most consequential commercial reaction is the hydrogenation of adiponitrile (ADN) to 1,6-hexamethylenediamine (HMD), a precursor to Nylon 6,6. In 2016, world consumption of HMD was estimated to be 1.46 million metric tons and expected to grow to 1.63 million metric tons in 2021, at an average annual growth rate of 2.2% [6].

Commercially, HMD is manufactured by the hydrogenation of ADN, either under high temperature and pressure or, alternatively, under relatively low temperature and pressure conditions [3]. The high-temperature and pressure processes have been described as employing a packed bed reactor in the presence of excess hydrogen and anhydrous ammonia at temperatures between 85 and 185°C and pressure from 27.5 to 41.4 MPa, i.e., 4,000–6,000 psig [7, 8]. In the processes, reduced cobalt oxide or iron oxide is used as a catalyst. The catalyst is first calcined in air and then reduced in a mixture of hydrogen and ammonia, at a temperature ranging from 300 to 600°C, in place, before using it in the hydrogenation reaction. On the other hand, patents from Monsanto [9, 10] delineate the use of Raney® nickel catalyst in a bubble-column reactor. The process described in these patents uses low temperature (50–100°C) and low pressure (1.5–3.6 MPa, i.e., 200–500 psig) to carry out the reaction. Instead of ammonia, the process uses sodium hydroxide to maintain the catalyst activity and reduce the formation of by-products.

The desired product for the hydrogenation of ADN is HMD, a primary amine, but high boiler impurities including one or more secondary amines are also formed as deleterious by-products (Figure 5.1). Maximum level of impurities in polymer grade HMD is quite restrictive [50 ppm of 1,2-diaminocyclohexane, 10 ppm of 6-aminocapronitrile (ACN), 100 ppm of 2-(aminomethyl)-cyclopentylamine, 25 ppm of hexamethyleneimine, 100 ppm of tetrahydroazepine, and 50 ppm of NH_3] and requires strict control of by-product formation during synthesis and refining [11]. In the high-pressure and temperature process, copious amounts of ammonia are added to suppress the formation of secondary and tertiary amines by shifting the equilibrium condensation reactions back toward primary amine (see Section 5.3). Otherwise, inevitably, there is dimerization and/or oligomerization yielding secondary and potentially tertiary amines on the surface of the Fe or Co catalyst, which generally leads to the deactivation of the catalyst by the blockage of the active sites. Similarly, in the case of Raney® Ni catalyst, if the hydrogenation reaction is carried out in the absence of a base, rapid catalyst deactivation occurs [12].

A number of studies have been reported in the literature, where Raney® Co catalyst has been used advantageously for ADN hydrogenation. Although the activity of Raney® Ni catalyst is higher than that of Raney® Co, the latter is much more selective for the synthesis of HMD [3, 13, 14]. Raney® Ni catalyst is known to form cyclic imines and condensation products, which requires expensive separation and

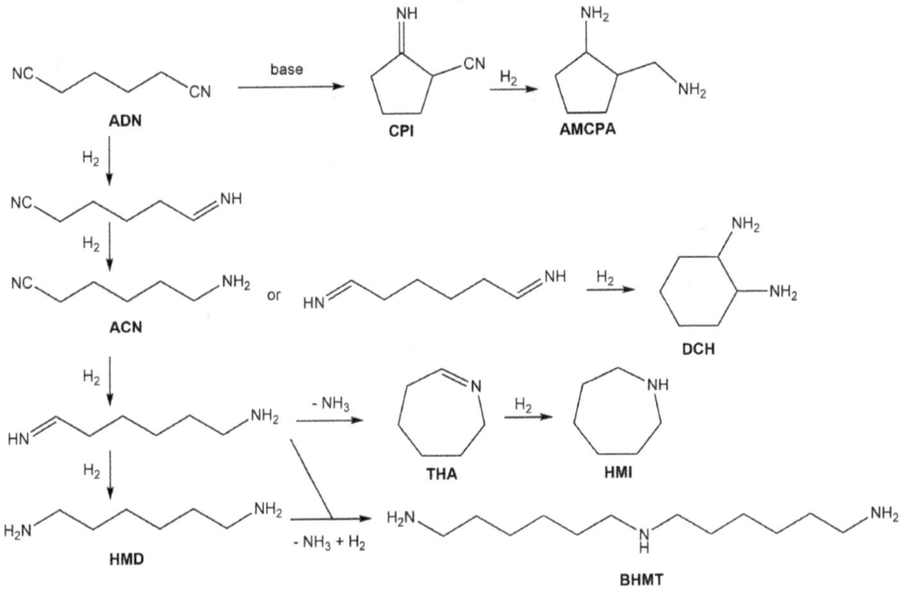

Figure 5.1: Reaction pathway for the formation of hexamethylenediamine and associated by-products by the hydrogenation of adiponitrile.

purification steps to deliver the purity necessary for producing Nylon 6,6 polymer. Lin et al. disclosed a high-pressure process for preparing HMD, where the latter is produced in a fixed-bed reactor by hydrogenating ADN in the presence of granular Cr and Ni promoted Raney® Co catalyst [15]. The reaction was carried out in the presence of at least 5 wt% ammonia (based on ADN) at a temperature and pressure in the range of 60–125°C and 0.03–34.5 MPa (50–5,000 psi), respectively. Sengupta et al. have shown that when Raney® Ni was replaced by Raney® Co in the low-temperature and pressure ADN hydrogenation process, the addition of sodium hydroxide was not necessary to maintain the activity, selectivity, and the life of the catalyst [16]. The reaction was carried out in a continuous stirred tank reactor (CSTR). Almost quantitative yield of HMD was obtained, with little or no catalyst deactivation, even after 13,242 g of ADN fed per gram of catalyst.

5.2.2 6-Aminocapronitrile synthesis

An exciting option for nitrile reduction is the partial hydrogenation of ADN to ACN because ACN can be used to produce caprolactam, the monomer of Nylon 6. Current global demand for caprolactam is approximately 5.2 million metric tons and expected to increase to nearly 6 million metric tons in 2021 at an average annual growth rate of 2.6% [17].

The production of Nylon 6 using ACN as a precursor (Figure 5.2) would be a much greener process compared to the current commercial production of caprolactam, as the latter process generates up to 2.3 kg of by-products per kilogram of caprolactam due to the use of hydroxylamine hydrosulfate and sulfuric acid [18]. In 2000, the then technical development manager for Royal DSM's fiber intermediates business, Ronald van der Stoel, commented, "This has been a nuisance for 40 years" [18].

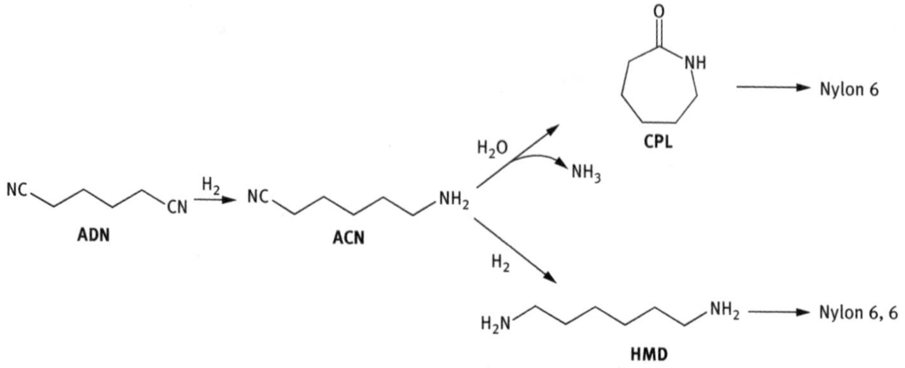

Figure 5.2: Synthetic routes to Nylon 6 and Nylon 6,6 via adiponitrile hydrogenation.

Due to the symmetrical nature of ADN, it is difficult to obtain higher than statistical distribution of ACN ($k1/k2 = 1$), unless a catalyst modifier or a chemoselective solvent is used, which would recognize and allow ADN to react and prohibit ACN from further surface reaction (Figure 5.3). This analysis employs apparent successive, first-order reactions.

Rigby patented the preparation of ACN using a high-pressure (13.8–20.7 MPa, i.e., 2,000–3,000 psi) batch hydrogenation process at 120°C, in the presence of anhydrous ammonia and Ni/alumina [19] and Co-promoted Raney® Fe and Co/kieselguhr [20] catalysts, where approximately 50–70% of theoretical yield of ACN was obtained. A greater than statistical distribution of ACN ($k1/k2 = 1.4–1.8$) has been reported by Allgeier using catalyst modifiers, such as tetrabutylammonium cyanide, tetraethylammonium fluoride, and tetramethylammonium hydroxide pentahydrate in conjunction with a reduced iron oxide catalyst, in the presence of ammonia as a solvent, in a batch process, at high pressure (31.0 MPa, i.e., 4,500 psig) and temperature (150°C) [21].

High selectivity of ACN (66% selectivity at 86% conversion) under low-temperature (80°C) and pressure (3 MPa) condition using Ni/SiO_2 catalyst, methanol, and caustic solution has also been reported in the literature [22]. On the other hand, Ziemecki obtained 89.5% 6-aminocaprontrile at 70% conversion of ADN in a batch autoclave, at a temperature of 70°C and under 3.4 MPa (500 psi) of hydrogen pressure, containing ADN, methanol, caustic solution, and Raney® Ni catalyst [23].

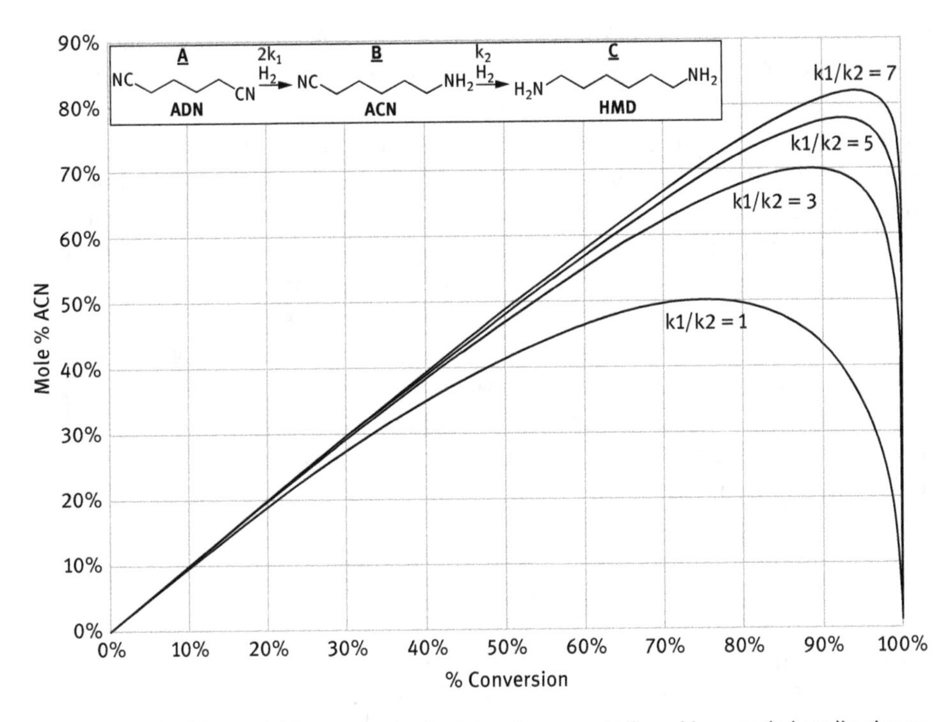

Figure 5.3: Selectivity model for the synthesis of 6-aminocapronitrile and hexamethylenediamine on the pathway to Nylon 6 and 6,6.

A plethora of selectivity modifiers juxtaposed with Raney® Ni, Co, and Fe have been reported in the literature to improve the yield of ACN. Formamide has been used as a selectivity modifier to boost the ACN selectivity beyond statistical distribution of ACN and HMD. A selectivity of ~85% ACN was obtained at 90% conversion of ADN, when the reaction was carried out in a batch reactor (BR) with ammonia as a solvent, Raney® Ni as a catalyst, and formamide as a selectivity modifier [24]. Similar result was obtained when the catalyst was pretreated before the reaction. Addition of *N*-methyl formamide, ammonium formate, urea, and ethyl formate also showed selectivity improvement [24]. Similar selectivity improvement was observed with a number of modifiers, including tetraethylammonium cyanate, tetramethylammonium hydroxide, tetraethylammonium thiocyanide, tetrabutylammonium thiocyanide, tetraethylammonium fluoride, when used with Raney® Ni and Co as a catalyst and ammonia as a solvent [25–28].

5.2.3 Acetonitrile hydrogenation

Although the hydrogenation of acetonitrile to its primary product, ethylamine, seems to be quite simple at the outset, the formation of diethylamine (secondary amine) and triethylamine (tertiary amine) makes the reaction mechanism more complicated (see Section 5.3). The reaction has been performed in the vapor phase, as well as in the liquid phase. Depending on the type and nature of the catalyst and the reaction medium, the final product distribution may include primary, secondary, or tertiary amine exclusively, or as a combination thereof. Verhaak et al. performed hydrogenation of acetonitrile in the vapor phase, in a fixed-bed reactor using supported nickel catalysts [29]. The study showed unequivocally that the nature of the support had a tremendous impact on amine selectivity. It was observed that, primarily, condensation products, such as diethyl- and triethylamine, were formed when acidic supports were used in heterogeneous nickel catalysts. However, high selectivity of ethylamine, a primary amine, was obtained when a basic support was used for a supported nickel catalyst. Based on the results obtained with a supported nickel catalyst having alkali-modified alumina support, a dual-site mechanism for the acetonitrile hydrogenation has been postulated [29].

5.2.4 Butyronitrile hydrogenation

Since the seminal work of Greenfield [30], time and time again, research investigators have found it to be intellectually expedient to use butyronitrile hydrogenation as a model reaction system because the reaction has been extensively studied, well defined, and elegant. Most of these studies are geared toward a fundamental understanding of the reaction mechanism of nitrile hydrogenation or investigating the activity, selectivity, and stability of a new catalytic system. However, it must be underscored that in addition to butyronitrile hydrogenation to butylamines being of great interest to the research community as a model reaction system, butylamines are also extremely important industrial chemicals in the overall alkylamines family, as discussed in an IHS Report [31]. The global consumption of butylamines is projected to be 80,000 metric tons by 2018. Butylamines are well-established intermediates for the manufacture of a plethora of rubber-processing chemicals, agrichemicals, pharmaceuticals, plasticizers, phase transfer catalysts, emulsifiers, adhesives, and dyes. The majority of butylamines industrial manufacturing processes involve the amination of alcohols. Additionally, reductive amination of aldehydes, Ritter reaction, amination of isobutylene, and nitrile hydrogenation are also practiced industrially to produce primary, secondary, and tertiary butylamines. Some of these processes are equilibrium limited, others require extensive purification and separation steps, and many of them are costly processes. As a result, many industrial research investigations have focused on exploring safer, efficacious, and cost-advantaged butyronitrile hydrogenation processes [31].

Hydrogenation of butyronitrile is a great example of reactions, where the types of active metallic species (precious vs. base metals), promoter, and solvent have a profound effect on the activity and chemoselectivity of supported and unsupported metal catalysts. The primary reaction pathway for the hydrogenation of butyronitrile proceeds through the formation of butylimine, which subsequently gets further hydrogenated to *n*-butylamine. However, the secondary reactions to form dibutylamine (secondary amine) and tributylamine (tertiary amine) can become more significant by the successive nucleophilic addition of butylamine to butylimine and *N*-butylidene-diamine (Schiff base), respectively (Figure 5.4). While Raney® Ni and Ru/SiO_2 catalysts predominantly produce *n*-butylamine as the primary product, Pt/SiO_2 and Pd/SiO_2 catalysts in *n*-butanol solvent formed mostly dibutylamine and minor amounts of tributylamine [32]. However, when tertiary amine is the desired product, the combination of catalyst and solvent may not be enough to provide the chemoselectivity required in an industrial process, when the catalyst is recycled multiple times. In most cases, the selectivity to tertiary amine drops over repeated use of the catalyst and requires washing of the catalyst in between each use. It has been reported by Vedage and Armor that the selectivity to tributylamine can be substantially enhanced by adding dichloroethane, as a selectivity modifier, to butyronitrile feed [33]. During the liquid-phase hydrogenation of butyronitrile, when 5 wt% Pd/Al_2O_3 was used as a catalyst, at an operating temperature and pressure of 125°C and 3.1 MPa (450 psi), respectively, it was observed that the selectivity to tributylamine dropped from 94% to 86%, over three consecutive runs with the same catalyst and in the absence of any solvent. In order to keep the conversion constant at ~85%, in each of the recycle runs, the time on stream had to be increased from cycle to cycle. During the fourth run, when 0.5 wt% dichloroethane was added to the neat butyronitrile feed, the selectivity to tributylamine could be reverted back to 96%, although at the expense of conversion of butyronitrile (70% vs. ~85%). A similar effect was observed at the addition of dichloroethane as a catalyst modifier, in the presence of isopropanol as a solvent [33].

Thomas et al. observed that the addition of iron and/or chromium to Raney® Ni catalyst increased the activity and Brunauer-Emmett-Teller (BET) surface area of the catalyst for the hydrogenation of butyronitrile, whereas it had little or no effect on the selectivity and CO chemisorption [35]. At lower hydrogen partial pressure (1.4 MPa, i.e. 206 psi), the order of reaction changed from half-order to first order in butyronitrile concentration as a function of temperature. Significant increase in rate was observed at higher hydrogen partial pressure, while the selectivity remained almost the same. Interestingly enough, the addition of a base, such as sodium hydroxide, had a significant influence on the formation of primary versus secondary and tertiary amine formation. The authors concluded that the increase in the selectivity of *n*-butylamine due to the addition of base suggested that the presence of base influenced the competitive adsorption of butylimine versus butylamine. A lower adsorption of butylamine would lead to the suppression of the formation of the condensation products.

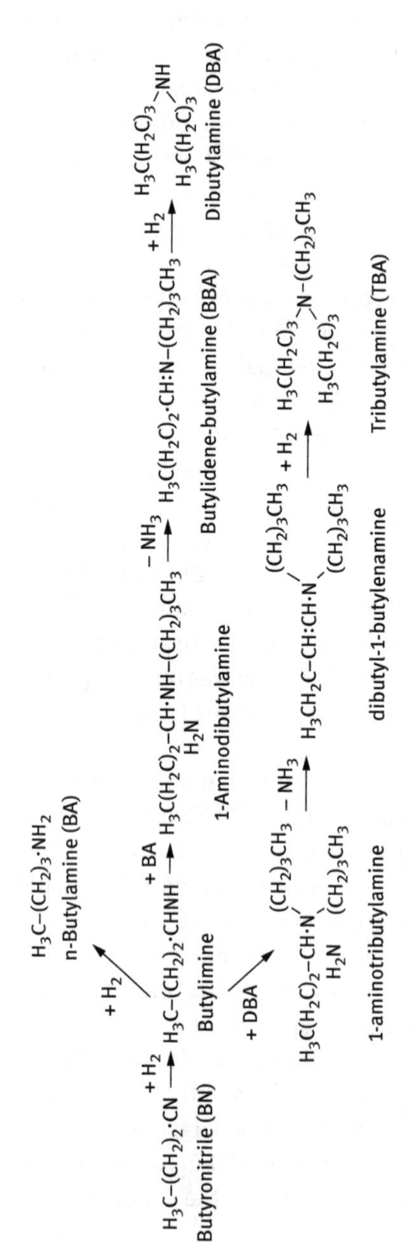

Figure 5.4: Reaction network of butyronitrile hydrogenation and coupling to amines [34].

5.2.5 Vitamin B1 (thiamine) synthesis

Thiamine, also called vitamin B1, is found in most foods, though mostly in small amounts. It is commercially available as thiamin chloride hydrochloride (known as thiamin hydrochloride) and the mononitrate. In 2017, the global demand of vitamin B1 market is estimated to be 8,500 MT [36]. According to market intelligence firm China Chemicals Market (CCM), vitamin B1 feed grade was $49,933 per ton in June 2016.

Vitamin B1 is primarily manufactured by chemical routes, although a few fermentation processes are also beginning to be practiced commercially. Two chemical routes are predominantly used to manufacture thiamine, namely, (1) the condensation of pyrimidine and thiazole rings and (2) the construction of thiazole ring on a preformed pyrimidine moiety [37].

The beneficial effect of a catalyst modifier to improve chemoselectivity is advantageously used in the synthesis of vitamin B1. 4-Amino-5-aminomethyl-2-methylpyrimidine (Grewe diamine) is a key intermediate for the synthesis of vitamin B1 and is prepared by the hydrogenation of 4-amino-5-cyano-2-methypyrimidine (also known as 5-cyano pyrimidine, pyrimidino nitrile, or pynitrile). Supported platinum and palladium catalysts and unsupported base-metal catalysts have been used to reduce pynitrile to Grewe diamine (primary amine) using nickel catalysts. Substantial amounts of secondary amine by-product were formed when supported Pd and Pt catalysts were used. Similar results were also obtained with nickel catalysts but the formation of secondary amines could be reduced to less than 5% by adding ammonia in the reaction mixture [38]. Degischer and Roessler reported 96.4% yield of the primary amine (Grewe diamine) and 1.9% of the secondary amine in the presence of Raney® Ni [39]. However, when using Raney® Ni catalyst pretreated with aqueous formaldehyde at ambient temperature, a 99.6% selectivity of the primary amine (<0.1% selectivity of the secondary amine) was obtained when the pynitrile was reduced under the same operating conditions. Comparable yield improvement of the primary amine was obtained when Raney® Ni was pretreated with carbon monoxide and Ni/SiO_2 catalyst was treated with formaldehyde. Ostgard et al. reported similar improvement in the selectivity of Grewe diamine, when activated Ni was pretreated with formaldehyde, carbon monoxide, acetone, or acetaldehyde, although the latter two modifiers were not as effective as the first two [40]. The decomposition and restructuring of the modifiers on the surface of the catalysts leading to the generation of more selective sites were suggested as a possible mechanism for the selectivity enhancement of the catalysts.

5.2.6 Isophorone diamine synthesis

Isophorone diamine (IPDA) is an important polymer intermediate used in the coatings and performance materials industries. Traditionally, it is used in polyurethanes, paints, curing agents, coatings, and varnishes [41, 42]. IPDA is produced from 3-cyano-3,5,5-trimethylcyclohexanone or isophorone nitrile (IPN), which is prepared by base-catalyzed addition of hydrogen cyanide to isophorone at 125–275°C [43]. The commercial IPDA process is a two-step process. In the first step, IPN is converted into ketimine 3-cyano-3,5,5-trimethyl-cyclohexaneimine, using an acidic catalyst, such as silica, alumina, titania, organic ion exchange resins, zeolites, and supported heteropolyacids. In the second step, the latter is hydrogenated in the presence of ammonia, hydrogen, and a fixed-bed Raney® cobalt catalyst [44, 45]. The production of IPDA is carried out in continuous fixed-bed reactors at high temperature (80°–200°C) and pressure (8–30 MPa) [46]. Interestingly enough, Haas et al. reported that the yield of IPDA improved from 89.7% to 95.7% when quaternary ammonium hydroxide base, such as tetramethyl ammonium hydroxide, was added to the hydrogenation feed containing ketimine, ammonia, and methanol [47]. The reaction was carried out in a continuous fixed-bed reactor containing Raney® Co catalyst at a temperature of 100°C and pressure of 6 MPa.

The fixed-bed Raney® Co catalyst used in the hydrogenation reaction was prepared by adding metallic cobalt as a binder to the cobalt–aluminum alloy to provide the necessary crush strength and attrition resistance of the formed catalyst [45]. An alternative method for the preparation of the fixed-bed Raney® Co catalyst for ketimine reduction has been described by Sauer et al., wherein only a thin outer layer of 0.1–2.0 mm in thickness of the cobalt–aluminum alloy is activated to prepare the formed catalyst [48]. The disadvantages of the above two types of formed Co catalysts are that they have very high bulk densities and a significant portion of the catalysts, in the form of binder and/or cobalt–aluminum alloy, remain inert in the hydrogenation process. To avoid these shortcomings and to improve internal mass transfer limitation in the formed catalyst particles, a hollow spherical rendition of the formed Raney® Co catalyst was patented by Ostgard et al. [42, 48]. The latter type of catalyst has the advantage of having a lower bulk density than the other types of fixed-bed Raney® Co catalyst, used in the ketimine reduction, while using less amount of cobalt for similar or higher hydrogenation activity.

Last but not the least, process intensification by combining reductive amination and nitrile hydrogenation in a single step has been demonstrated by Haas et al. [49]. Yield of up to 88.7% IPDA, based on IPN, was obtained when a mixture of IPN, methanol, and ammonia in the ratio of 1:3.2:2.2 was hydrogenated in a continuous fixed-bed reactor containing 50% Co on silicate support in the form of extrudates at a temperature of 120°C and pressure of 6 MPa. Five-percent Ru on γ-Al_2O_3 extrudates also afforded 80% yield of IPDA [49].

5.2.7 Hydrogenation of benzonitrile

The synthesis of benzylamine was first reported by Rudolf Leuckart in 1885 when, serendipitously, he prepared benzylamine by heating benzaldehyde with formamide, while trying to prepare benzylidenediformamide, $C_6H_5CH(NHCHO)_2$ [50]. However, benzylamine, a versatile intermediate and building block for producing active pharmaceutical ingredients (API), corrosion inhibitors, military explosives, dyestuffs, synthetic resins, and rocket propellants, is currently commercially manufactured by the hydrogenation of benzonitrile or amination of benzaldehyde or benzyl chloride [51]. Of all the reaction pathways, the most convenient, atom efficient, and economically viable route to prepare benzylamine is the catalytic hydrogenation of benzonitrile [52, 53].

Like many nitrile hydrogenation reactions, the challenge one faces for the catalytic hydrogenation of benzonitrile is the loss of selectivity of the primary amine due to the formation of secondary and, potentially, tertiary amine by-products (Figure 5.5). Copious amount of ammonia is usually added to suppress the formation of secondary and tertiary amines. It has been shown that Raney® nickel and supported rhodium catalysts provide the highest selectivity toward the primary amine, whereas secondary amine seems to be the primary product in the presence of palladium and platinum catalysts, even though a large excess of ammonia was added [52, 53].

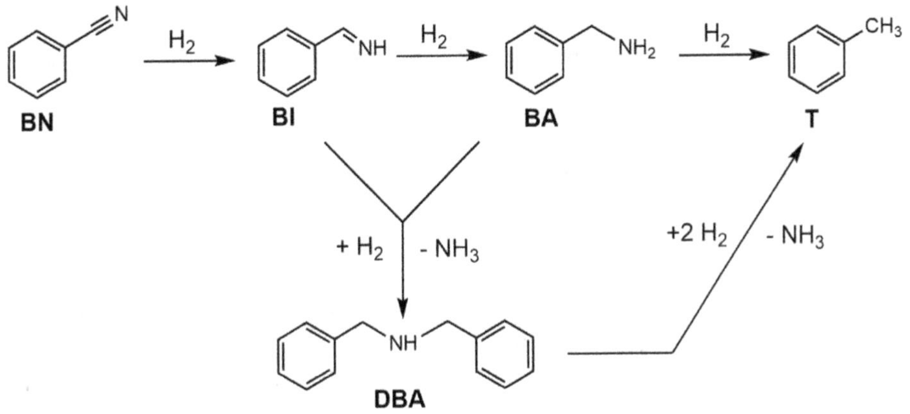

Figure 5.5: Potential reaction pathways in the hydrogenation of benzonitrile [52].

The quest for high selectivity of primary amine at high conversion of aromatic nitrile still intrigues researchers. Recently, Saito et al. reported the selective synthesis of benzylamine by the hydrogenation of benzonitrile in the presence of poly (dimethylsilane)-supported Pd catalysts with different second supports (DMPSi-Pd/support), such as silica in a continuous fixed-bed reactor [53]. 1-Propanol was used as the primary solvent and water was added as a cosolvent (ratio of 4:1) with benzonitrile substrate (0.2 M) as the feed (flowrate of 0.2 mL.min^{-1}). Reactions were performed for 18 h, time on stream, at 60°C, and 5.0 MPa hydrogen partial pressure. Interestingly enough, the addition of 1.5 equivalents of HCl led to quantitative yield of benzylamine even under such mild operating conditions, even though earlier studies have shown that supported Pd catalysts preferentially produce secondary amines. Furthermore, the life of the catalyst was demonstrated for 300 h, time on stream. The product was obtained in quantitative yield (TON = 10,078) with little or no loss of Pd or leaching of silica support.

While the work of Saito et al. might be pathbreaking from the point of view of using a Pd catalyst under mild operating conditions to obtain quantitative yield of benzylamine, it still suffers from the fact that 1.5 equivalents of HCl is necessary for the efficacy of the process. From a commercial point of view, HCl is corrosive and requires expensive material of construction and it will require recycling and reusing HCl to make the process "green" and economically viable. On the other hand, Hegedűs and Máthé delineated an interesting liquid-phase, selective, heterogeneous, catalytic hydrogenation process, whereby they could achieve an isolated yield (85–90%) of benzylamine at complete conversion under mild operating conditions, as well (30°C and 0.6 MPa) [52]. Apparently, the uniqueness of their process is the use of a biphasic solvent system (water/dichloromethane) over Pd/C catalyst, in acidic medium by adding sodium dihydrogen phosphate (NaH_2PO_4) as an additive. The authors claim that very high purity of benzylamine (>99%) could be obtained without applying any special purification technique and assert that the process can be easily and economically scaled up to produce commercial quantities of primary benzylamine.

5.2.8 Nitrile hydrogenation in pharmaceuticals

As a highly atom-efficient transformation, nitrile hydrogenation has been utilized in production of pharmaceuticals. Tranexamic acid is an antifibrinolytic drug identified by the World Health Organization as an essential medicine and which employs catalytic nitrile hydrogenation [54]. Sera and coworkers developed a multi-kilogram scale synthesis of a DPP-4 inhibitor for treatment of diabetes employing a Raney® Ni-catalyzed nitrile hydrogenation [55]. Additionally, Saravanan and coworkers developed an alternate synthesis of venlafaxine with reduced cost and improved quality; venlafaxine is indicated for the treatment of depression and anxiety [56]. The synthesis employed Raney® Ni-catalyzed nitrile hydrogenation in acetic acid solvent. The challenges of nitrile hydrogenation for the production of API derive from functional group diversity and solubility concerns associated with this class of substrates.

5.3 Mechanism and process parameters

5.3.1 Intermediates

As early as 1905, Sabatier and Senderens [2] studied the hydrogenation of nitriles and made cursory suggestions of the mechanism involving stepwise addition of H_2 to form intermediate aldimine (i.e., primary imine), followed by primary amine. With time, a more sophisticated view of the nitrile hydrogenation mechanism emerged accounting for primary products and by-products, primarily secondary, and tertiary amines. The oft-cited von Braun mechanism [57] with the Greenfield modification [30] (Greenfield proposed the intermediacy of the enamine for tertiary amine formation) is widely accepted as consistent with most observed intermediates and products but, notably, does not incorporate any surface reactions (Figure 5.6).

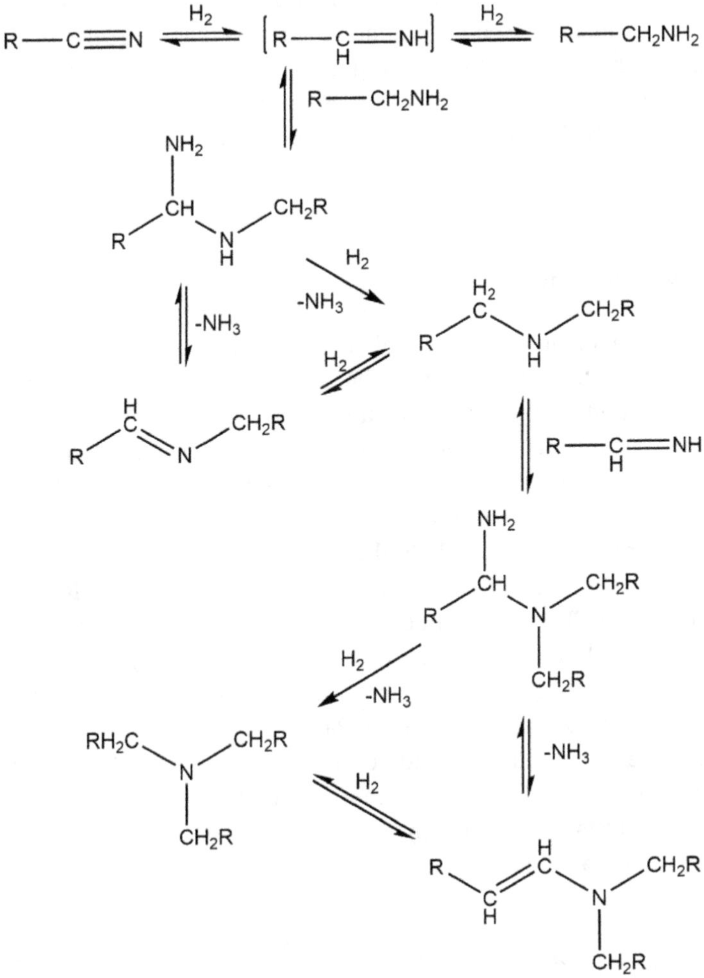

$R-C\equiv N \xrightleftharpoons{H_2} [R-CH=NH] \xrightleftharpoons{H_2} R-CH_2NH_2$

$[R-CH=NH] + R-CH_2NH_2 \rightleftharpoons R-CH(NH_2)-NH-CH_2R$

$R-CH(NH_2)-NH-CH_2R \xrightleftharpoons{-NH_3} R-CH=N-CH_2R$

$R-CH(NH_2)-NH-CH_2R \xrightarrow{H_2,\ -NH_3} R-CH_2-NH-CH_2R$

$R-CH=N-CH_2R \xrightleftharpoons{H_2} R-CH_2-NH-CH_2R$

$R-CH_2-NH-CH_2R + R-CH=NH \rightleftharpoons R-CH(NH_2)-N(CH_2R)_2$

$R-CH(NH_2)-N(CH_2R)_2 \xrightarrow{H_2,\ -NH_3} RH_2C-N(CH_2R)_2$

$R-CH(NH_2)-N(CH_2R)_2 \xrightleftharpoons{-NH_3} R-CH=CH-N(CH_2R)_2$

$R-CH=CH-N(CH_2R)_2 \xrightleftharpoons{H_2} RH_2C-N(CH_2R)_2$

Figure 5.6: Nitrile hydrogenation mechanism of von Braun [57] with modification by Greenfield [30].

Krupka and Pasek [4] provided an authoritative review of the mechanism of nitrile hydrogenation by heterogeneous catalysts and readers are directed there for a detailed discussion. The intermediacy of the aldimine has not been directly observed, in spite of multiple attempts, including very sensitive GC–MS evaluation [58]. We have, additionally, attempted to observe the intermediate utilizing *in-situ* infrared spectroscopy of liquid-phase ADN hydrogenation, to no avail. The absence of the observation could be attributed to the high reactivity of the intermediate aldimine at the catalyst surface or it may be that the surface bound intermediate is not the classic aldimine described in the von Braun proposal [4]. Krupka and Pasek [4] describe the thesis work of Z. Severa seeking to provide evidence of aldimine in the homogenous (liquid) phase. The addition of solid acids to liquid-phase Raney-type nickel catalysts to enhance condensation reactions of the putative aldimine led to no change in selectivity suggesting that if the aldimine exists it does not desorb from the catalyst surface. Circumstantial evidence for the intermediacy of the primary aldimine was obtained by Singh and coworkers in the hydrogenation of benzonitrile in the presence of water, which yielded benzaldehyde [59].

5.3.2 Surface bound species

Fouilloux and De Bellon made an early mechanistic proposal inspired by organometallic chemistry concepts but without direct characterization of surface bound intermediates [3]. Huang and Sachtler [60, 61] studied the deuteration of acetonitrile and observed significant incorporation of proton on the reduced nitrogen, particularly for metals known to form stable M=N double bonds (e.g., Ru). They inferred that the rate-limiting product desorption step involved concerted hydrogen transfer from the CH_3– group of acetonitrile to the surface Ru nitrene (Figure 5.7) and that the von Braun assumption of successive addition of D_2 to the nitrile and then the aldimine to form $CH_3CD_2ND_2$ must not be valid. The stabilization of various surface intermediates should depend on the identity of the catalytic metal sites. Scharringer et al. observed direct evidence for Co nitrene binding mode using inelastic neutron scattering for acetonitrile over Raney® Co catalyst [62]. They also observed correspondingly high selectivity to primary amine but lack of H/D exchange in the hydrogenation of CD_3CN, in contrast to Huang and Sachtler's observations over Ru catalysts. These differences help explain preferred reactivity of Co catalysts in the formation of primary amines [63, 64].

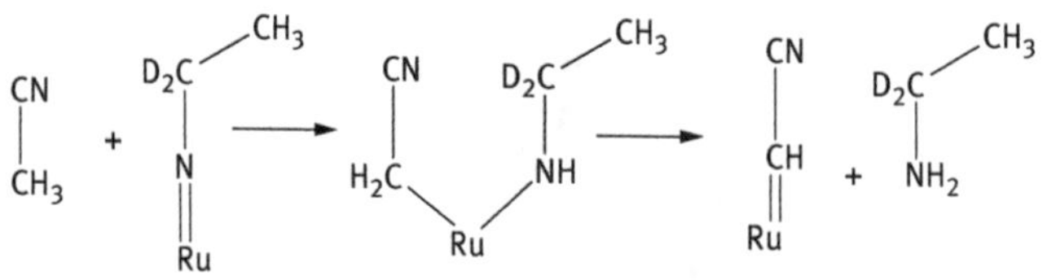

Figure 5.7: Hydrogen transfer in Ru-catalyzed nitrile reduction [60].

Coq et al. provided infrared spectroscopic evidence of Ni-C bound complexes of the type, acimidoyl and aminomethylcarbene, in the adsorption of CD_3CN on supported Ni containing catalysts [65]. They described these surface complexes as intermediates in the formation of secondary amines (Figure 5.8).

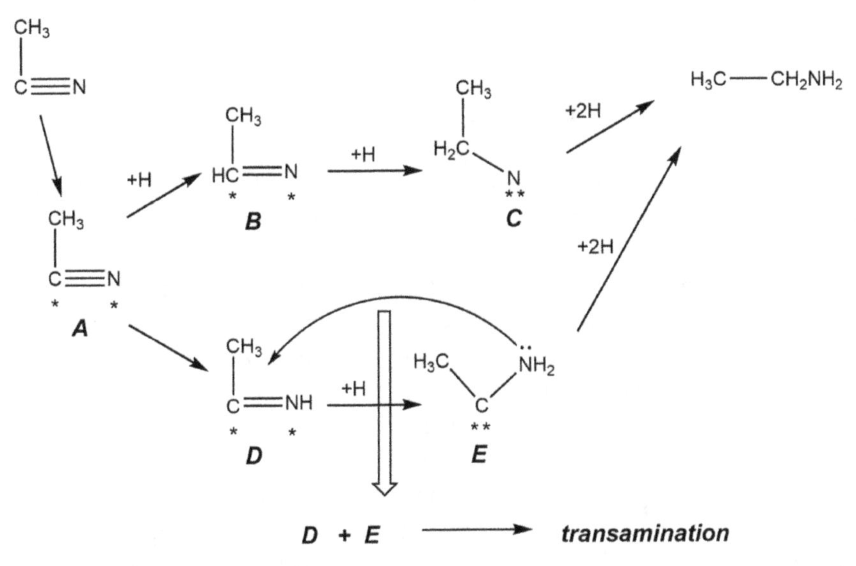

Figure 5.8: Proposed mechanism of nitrile hydrogenation and critical role of surface intermediates in product and by-product formation [65]. The symbol * indicates a surface site.

Ou et al. [66] applied infrared reflection absorption spectroscopy to characterize the η^2-binding of CN species over Pt surfaces, which enhances the likelihood of nucleophilic attack at the α-carbon. The preponderance of the literature precedent concerning surface bound intermediates in nitrile hydrogenation supports the view that the von Braun mechanism, while conceptually useful, is an over-simplification of the hydrogenation of nitriles over heterogeneous catalysts. The diversity of surface bound structures on various metallic surfaces may explain differences in selectivity to primary versus secondary and tertiary amines.

5.3.3 Secondary and tertiary amine formation

In keeping with the above discussion of the structure of surface bound intermediates relative to putative desorbed aldimines, multiple authors have concluded that the reactions forming secondary and tertiary amines during nitrile hydrogenation are metal-surface catalyzed reactions and do not occur exclusively in the liquid-phase or on remote acid sites [4]. Huang and Sachtler [58] demonstrated that butyronitrile hydrogenation in the liquid and gas phase exhibited essentially no difference in selectivity toward condensation reactions to form secondary amines for multiple metals, concluding that the reactive intermediates do not desorb from the surface and react in the homogeneous phase but, rather, that coupling reactions occur on the catalyst surface. Coq et al. [65] argued that formation of secondary amine originates from the bimolecular reaction of two surface bound intermediates and Roessler [67] and, subsequently, Ostgard [40, 68] demonstrated that catalyst poisoning to break-up ensembles of surface sites associated with such bimolecular surface reactions led to improvements in primary amine selectivity, all in agreement with Huang and Sachtler [58]. Still, it may be acknowledged that addition of exogenous amine will lead to asymmetric diamines from coupling reactions, e.g., pentylamine addition to butyronitrile hydrogenation led to the unsymmetrical diamine by-product [58]. Accordingly, exogenous amine can participate in coupling reactions directly reacting with surface intermediates or by reacting to become surface-bound intermediates, themselves. Volf and Pasek [14] described evidence that secondary amine formation for dodecanenitrile hydrogenation proceeds only through the dodecylidenedodecylamine (secondary imine), which accumulates in the reaction, and not through the 1-aminodidodecylamine (hemi-aminal), which the von Braun mechanism allows but which was not detected.

The selectivity of nitrile hydrogenation to primary, secondary and tertiary amines has a strong dependence upon the identity of the metal. Huang and Sachtler compared a series of M/Na-Y zeolite supported catalysts in the liquid phase hydrogenation of butyronitrile, Table 5.1 [58]. With Na/Y support, Ru provided high selectivity to primary amines, followed by Rh and Ni. Pd and Pt were particularly selective to secondary amines. Others have demonstrated even higher selectivity to tributylamine using Pd and Pt catalysts under different conditions [30]. (See discussion below on the influence of reaction parameters.) Notably absent was a cobalt variant, as cobalt is known to be the most selective toward primary amine products [14]. Schärringer et al. studied the mechanism of liquid-phase acetonitrile and butyronitrile hydrogenation over Raney® Co catalyst [69]. Acetonitrile was observed to hydrogenate more rapidly than butyronitrile in competition experiments and butylamine was found to be more rapid at forming coupling products on the path to secondary amines than shorter chain amines. Each of these observations correlated with differences in inductive effects for butyl versus ethyl or methyl. Further, these authors ascribed a two-site mechanism, having observed negligible dependence of nitrile hydrogenation rate on the concentration of added primary amine. The addition of primary amine to the reactions did enhance secondary imine formation rates, consistent with the conclusion of a separate active site.

As with secondary amines, there has been a discussion in the literature on the mechanism of formation of tertiary amines (Figure 5.9) [4]. The von Braun mechanism allows for hydrogenolysis of 1-aminotrialkylamine intermediates but as described in Krupka and Pasek's review [4], and elsewhere [70], tertiary amines are not formed in

Table 5.1: Liquid phase hydrogenation of butyronitrile over M/NaY catalysts [58] *BA*, Butylamine; *BBA*, butylidine-butylamine; *DBA*, dibutylamine; *TBA*, tributylamine.

	TOS	Conversion	Selectivity (mol%)			
Catalyst	**(h)**	**(%)**	**BA**	**BBA**	**DBA**	**TBA**
Ru/NaY	8	89.2	67.9	22.8	9.2	0.1
Rh/NaY	2	93.8	44.2	4.5	51.0	0.2
Ni/NaY	3	99.8	23.5	0.3	61.2	15.0
Pd/NaY	7	89.9	3.6	0.1	94.8	1.4

benzonitrile hydrogenation, suggesting that the elimination mechanism to form intermediate enamine structures is more likely for alkylnitriles. Notably, the mechanism is prohibited by the aromaticity of the phenyl ring (the absence of a β-H) in the case of benzonitrile, consistent with the observation of no tertiary amine formation in that case (Figure 5.9) [70].

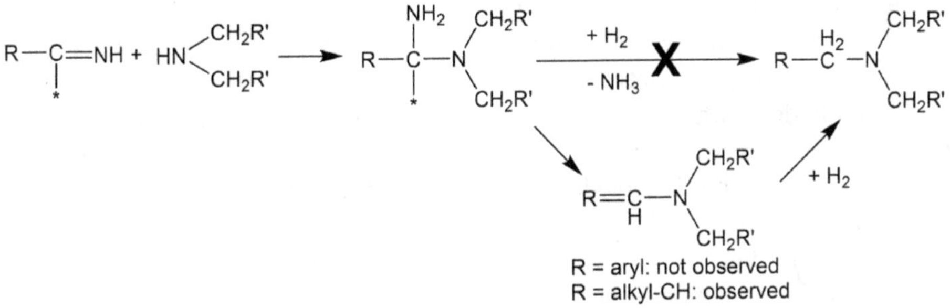

Figure 5.9: Proposed mechanism of tertiary amine formation via the intermediacy of enamine intermediates. The symbol * indicates H or bond to metallic surface.

Huang and Sachtler [58] provided the first direct evidence of the intermediacy of enamines in the formation of tertiary amines and noted the short-lived nature in the case of alkylvinyl-dialkylamine intermediates derived from butyronitrile hydrogenation but the enhanced stability in arene-stabilized vinyl-dialkylamines, derived from benzylcyanide ($PhCH{=}CHNR_2$) hydrogenation. Recently, Krupka et al. [71] demonstrated that even in the absence of β-H, tertiary amines could form from alkyl nitriles. Specifically, trimethylacetonitrile hydrogenation over Pd/C in the presence of diethylamine led to diethylneopentylamine (tertiary amine), though it is not feasible to form the corresponding enamine in this structure. The authors conclude that reactive surface intermediates are involved and that geminal elimination of ammonia can lead to a surface bound intermediate that can be hydrogenated to the tertiary amine.

Hydrogenation of pure trimethylacetonitrile and pure aryl nitriles did not lead to tertiary amines and Krupka et al. suggested that steric and electronic factors inhibit those products, even though formation of a surface bound aminocarbene is conceptually feasible [71].

Considering all the available literature evidence, Krupka and Pasek presented a unified mechanistic proposal of nitrile hydrogenation over metallic heterogeneous catalyst surfaces (Figure 5.10) [4]. As with any mechanism that seeks to be very general, some observations from the literature on specific systems are not accounted for by the mechanism. In this case, it's notable that the observation of secondary imine formation en route to secondary amines and enamine formation en route to tertiary amines [58] is not accounted for in the Krupka/Pasek mechanism nor is the influence of the organic moieties connected to the nitrile (e.g., aryl versus alkyl). Nonetheless, the unified mechanism accounts for the differences in selectivity observed for different metals and serves as a useful guide for further research in the field.

5.3.4 Dinitrile deactivation mechanism

A special case for mechanistic consideration is the deactivation of dinitrile hydrogenation catalysts, the hydrogenation of ADN to HMD being the most commercially significant case. Catalyst deactivation in commercial processes represents a significant factor in determining profitability. A systematic catalyst deactivation study was carried out for Raney® Ni catalyst by Allgeier and Duch for the hydrogenation of different classes of nitrile moieties (mono- and dinitriles and long and short-chain nitriles and dinitriles) [12, 72].

All α,Ω-dinitriles deactivated the catalyst, while no mononitriles deactivated the catalyst, except 6-aminohexanenitrile (Figure 5.11). Based on reactivity trends and electron spectroscopy for chemical analysis of fresh and used Raney® Ni catalysts, they concluded that the catalyst deactivated due to the formation and deposition of polyamines on the surface of the catalyst (Figure 5.12). Polyamine formation took place on the surface of the catalyst due to condensation reactions forming secondary amines in the absence of a base or ammonia. Larger chain-length molecules deactivated the catalyst more rapidly as they generated higher molecular weight polyamines for a given number of condensation events. It was postulated in the study that while ammonia prevented the formation of dimers and oligomers on the surface of the catalyst by reversing the equilibrium reaction toward the primary amine, the presence of an alkali hydroxide base blocked the adsorption of amine moieties on the surface of the catalyst via competitive adsorption of OH^- ions on the catalytic active sites [12]. It was shown that very little NaOH (0.2% w/w) was necessary to maintain the activity of the catalyst for a long duration of time, while a large excess of NH_3 (60% w/w) was necessary to do the same. This is a significant departure from a mechanism suggested by Verhaak et al. for the deactivation of Ni/SiO_2 catalyst,

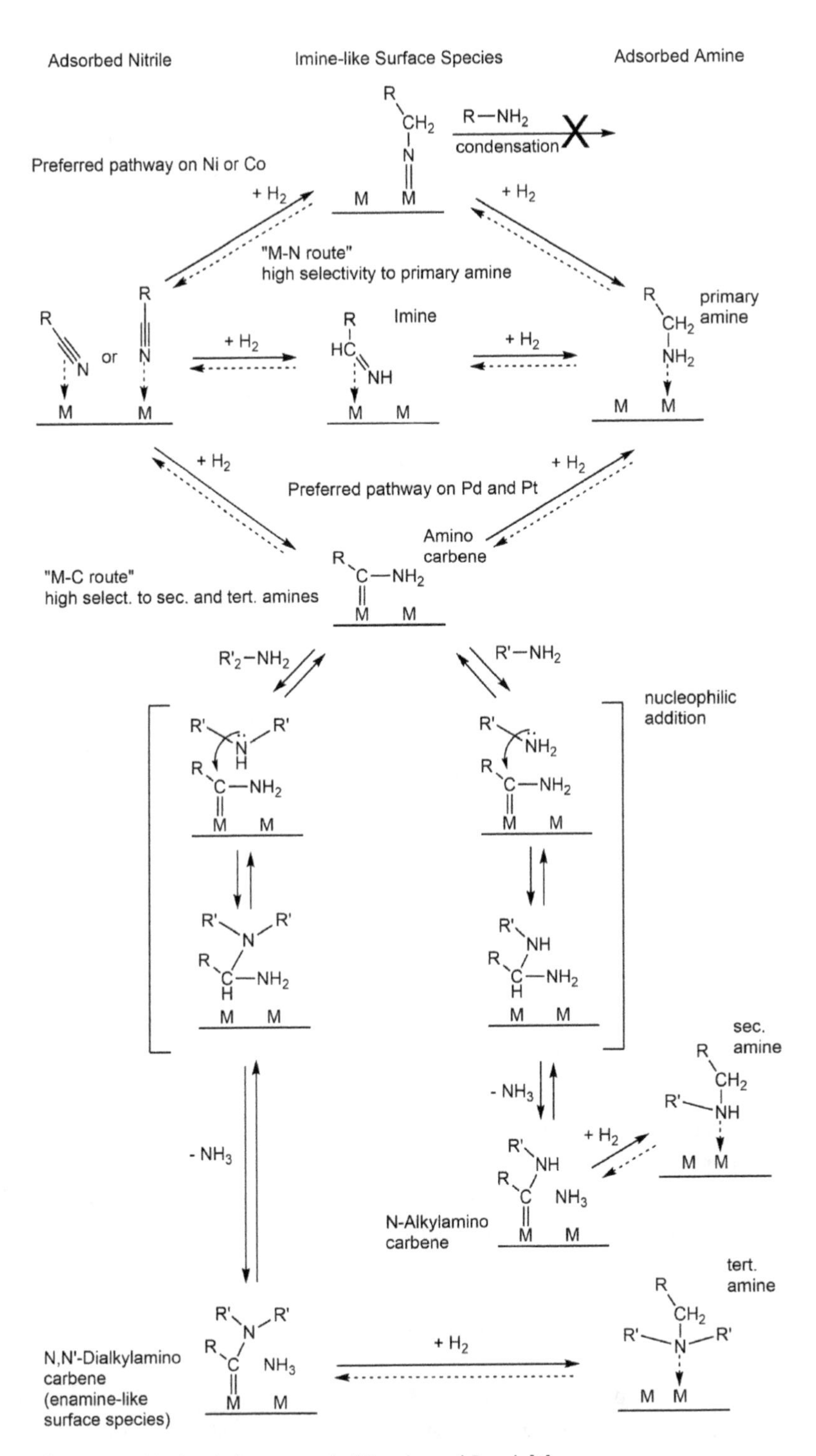

Figure 5.10: Mechanistic proposal of Krupka and Pasek [4].

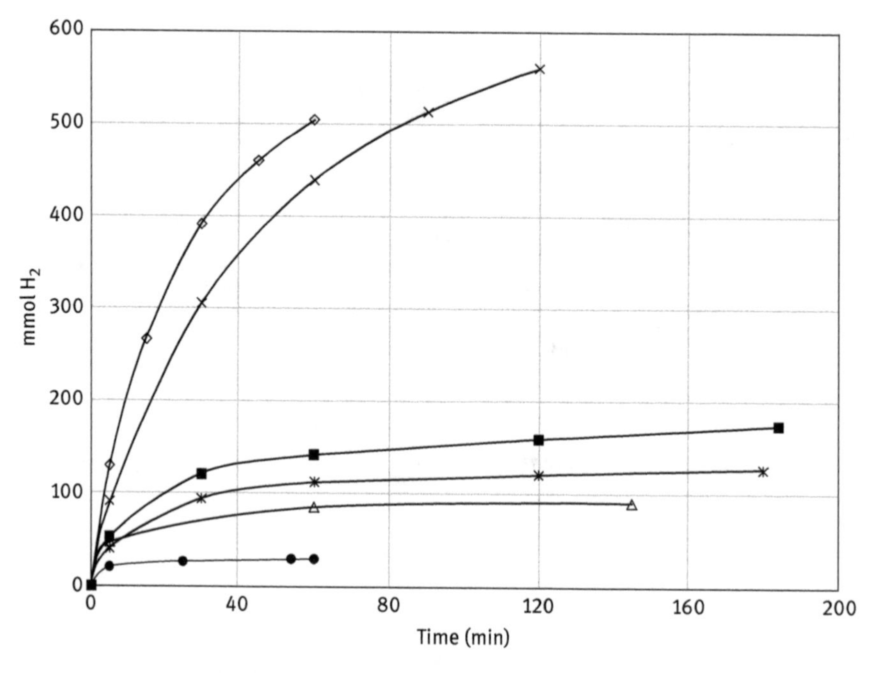

Figure 5.11: Reaction progress and deactivation for a series of nitrile hydrogenation reactions at 6 M nitrile in methanol (i.e., 3 M dinitrile). ◇: Butanenitrile, ×: hexanenitrile, ■: 1,5-pentanedinitrile, ⋇: 6-aminohexanenitrile, △: 1,6-hexanedinitrile, ●: 1,12-dodecanedinitrile. Reaction parameters: 75°C, 3.45 MPa H_2, stirred batch reactor.

Figure 5.12: Polyamine formation during hydrogenation nitrile.

during the hydrogenation of acetonitrile in the vapor phase in a fixed-bed reactor at 125°C and atmospheric pressure [29]. That study concluded that the root cause of the deactivation of Ni/SiO_2 catalyst was a cracking decomposition of acetonitrile, leading to the formation of nickel carbide on the surface of the catalyst.

5.3.5 Homogeneous catalysis mechanism

A description of the development and scope of homogeneous catalysts for nitrile hydrogenation is presented below; here, we focus on mechanistic studies specific to homogeneous catalysts. For the family of Ru hydride catalysts, Grey et al. described

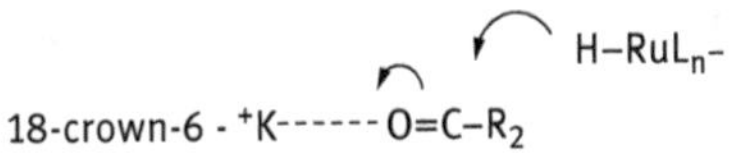

18-crown-6 · ^{+}K------$O{=}C{-}R_2$ H–RuL_n^-

Figure 5.13: Crown ether-promoted reaction rate of ketone hydrogenation (analog of nitrile hydrogenation) suggests that hydride transfer is in the rate-determining step [73].

seminal work, in 1981, on the hydrogenation of nitriles by a homogeneous Ru phosphine complex, invoking a hydride-transfer mechanism [73]. They observed reaction rate enhancement, of analogous ketone hydrogenation, upon the addition of crown ether, to complex the alkali counter ion of the negatively charged Ru complex (Figure 5.13), and concluded that the rate-limiting step did, indeed, involve hydride transfer, i.e., heterolytic activation of H_2 to a Lewis acid activated substrate.

Following groundwork on ligand-assisted, heterolytic activation of hydrogen by Ru–amido complexes [74, 75], Hidai, in 2002, described Ru–amido complexes as catalysts for benzonitrile hydrogenation and noted different product distribution when compared to Ru–phosphine-only complexes (Figure 5.14) [76]. Specifically, the main product with Ru–amido complexes was benzylamine but in the absence of amido ligand, the reaction favored coupling reactions to benzylidenebenzylamine. It is notable that higher rates were observed in the presence of strong base, which

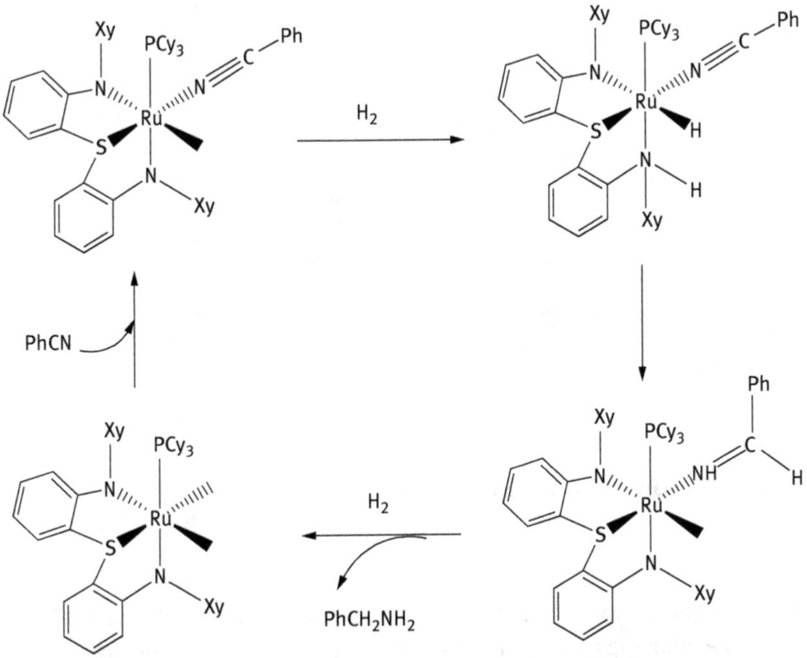

Figure 5.14: Heterolytic activation of hydrogen in Ru–amido complexes from Hidai [76].

purportedly abstracted a proton from the Ru–hydride–amine complex and facilitated hydride transfer to the substrate.

In 2012, Milstein reported a Ru pincer complex for nitrile partial reduction coupled with amine attack to selectively form imines [77] and proposed ligand-cooperativity and heterolytic activation of hydrogen by a different mechanism than Hidai [76]. Rather than localizing a proton on the ligand nitrogen atom, they proposed involvement of a ligand C–H bond and dearomatization of a pyridine ring in the ligand (Figure 5.15) [77]. Subsequent studies supported the mechanism with both nuclear magnetic resonance (NMR) evidence [78] and density functional theory (DFT)modeling [79].

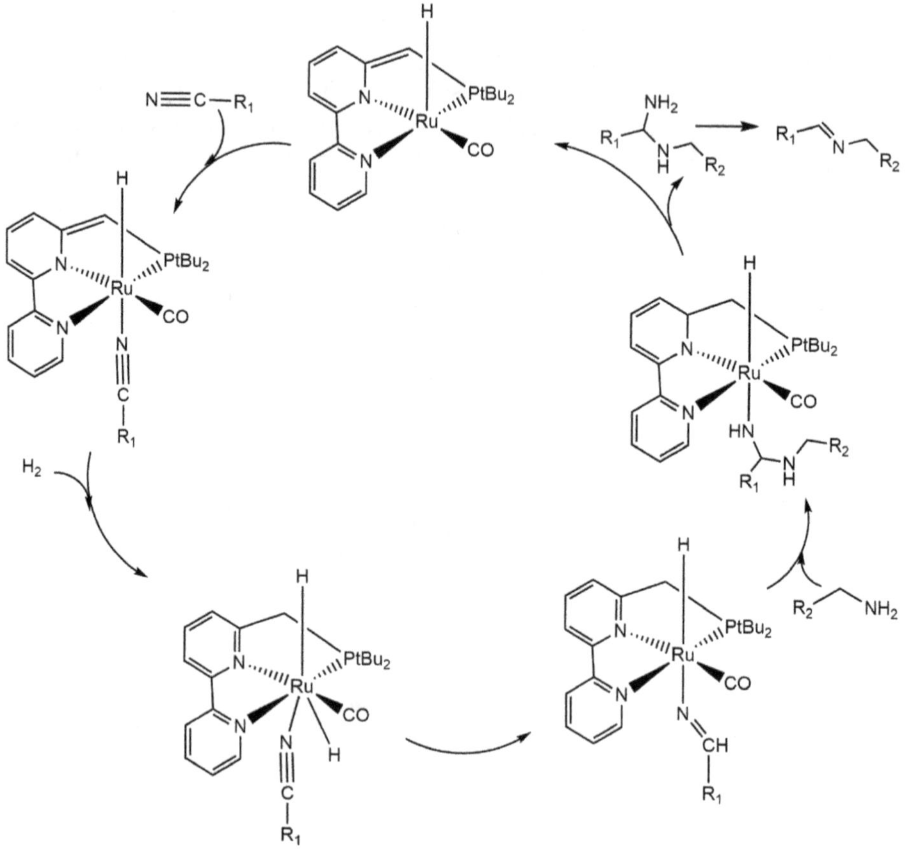

Figure 5.15: Ligand dearomatization during nitrile hydrogenation [77].

Beller and coworkers demonstrated the first homogeneous Fe catalyst for nitrile hydrogenation by employing a PNP-pincer ligand, wherein the selection of an aliphatic amine ligating group enabled heterolytic activation of H_2 by the catalyst (Figure 5.16)

[80]. Ligand cooperativity via an iron amide/amine cycle was, presumably, stabilized by the design of the ligand, placing the nitrogen as the middle ligating moiety relative to the Hidai example [76], where the amine was terminal. NMR studies focused on the ^{1}H hydride region and ^{31}P signal. When taken along with DFT modeling, the data supported a mechanism involving heterolytic activation of H_2.

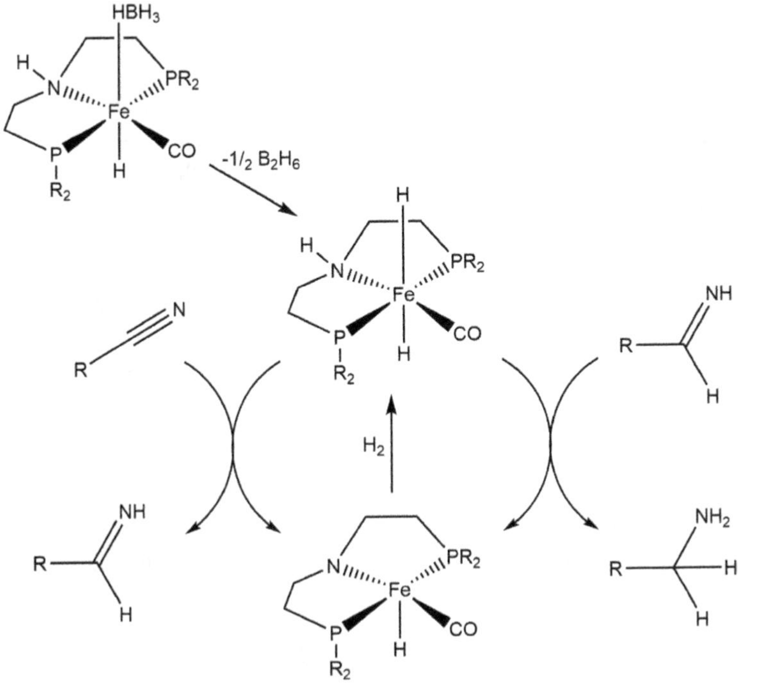

Figure 5.16: Homogeneous iron catalyst for nitrile hydrogenation [80].

Notably, the same ligand utilized in these Fe complexes has been deployed successfully for Mn nitrile hydrogenation catalysts with a comparable reaction mechanism [81]. Milstein, cleverly, designed Co catalysts, with the potential for both C–H (dearomatization) and N–H (amide/amine)-based ligand cooperativity in the heterolytic activation of hydrogen [82] and Fout and coworkers recently employed a chelating bis(*N*-heterocyclic carbene) Co complex describing the influence of Lewis acid in promoting the reaction [83], perhaps inspired by the early work of Grey et al. [73].

5.3.6 Influence of reaction parameters

Whether via homogeneous or heterogeneous catalysts, the hydrogenation of nitriles represents an atom-efficient organic transformation with great value compared to classic organic synthesis reductions involving aluminum or boron hydride reagents [84]. Reactions may be optimized via careful selection of reaction and reactor parameters [14].

5.3.6.1 Ammonia

For the production of primary amines, the use of ammonia as an additive/solvent is a traditional option to increase selectivity. Consistent with both the von Braun mechanism and mechanisms invoking surface bound species, ammonia shifts the equilibrium associated with condensation reactions en route to secondary amines [85]. Yields greater than 90% are readily attainable over Raney® Ni in the presence of 30 equivalents [86]. Ammonia has also been employed in temperature control schemes for plant designs [87].

5.3.6.2 Strong base

Alkali metal hydroxides are known to enhance the selectivity of nitrile hydrogenation to primary amines at much lower molar composition compared to ammonia and with good generality for a variety of noble and base-metal catalysts, Table 5.2 [3, 88]. In some cases, the alkali metal hydroxide is present at <0.5 g.g catalyst^{-1} and 0.002 equivalent compared to nitrile [63], whereas ammonia is utilized at 2–30 equivalent compared to nitrile [3]. For Raney® Co catalysts, LiOH was reported as preferable to NaOH and KOH, as the latter two led to agglomeration of catalyst particles under identical operating conditions [63]. Side reactions of substrates with base are known

Table 5.2: Influence of base additives on primary amine yield.

Catalyst	Temp. (°C)	Additive	Nitrile	Time (h)	Yield
Rh hydroxide[a]	80	None	ADN	0.75	35
Rh hydroxide[a]	80–100	NaOH	ADN	1.3	59
Raney Ni[a]	80	None	ADN	–	52
Raney Ni[a]	80	NH_3	ADN	–	78
Raney Ni(Cr)[a]	80	NaOH	ADN	–	100
Raney Co[a]	120	None	ADN	–	74
Raney Co[a]	110	NH_3	ADN	–	94
Raney Co(Ni)[b]	100	None	DMAPN	2.1	<91 (9.3% secondary amine)
Raney Co(Ni)[b]	100	LiOH	DMAPN	2.3	<99 (0.6% secondary amine)

[a] *ADN*, Adiponitrile; product = 1,6-hexamethylenediamine. Data from Ref. [3].
[b] *DMAPN*, Dimethylaminopropionitrile; product = 3-(dimethylamino)propylamine [63].

and should be considered in utilization of strong base additives [3]. In one study, the alkali hydroxide competed for surface sites binding primary amine and the reduced surface concentration of primary amine correlated with a reduction in secondary amine formation [35].

5.3.6.3 Temperature

A study of cobalt catalysts for lauronitrile hydrogenation revealed that secondary amine formation was preferred at higher temperatures [14]. The apparent activation energy for the condensation of primary amine with imine was measured as 132 kJ.mol^{-1}, whereas for nitrile hydrogenation, the value was 52 kJ.mol^{-1}. Similar observations were made for other cobalt and nickel catalysts, suggesting that lower temperatures favor primary amine formation [14].

5.3.6.4 Hydrogen pressure

Reference to the von Braun mechanism suggests that increases in hydrogen pressure should correlate with increased selectivity to primary amines during nitrile hydrogenation, and indeed for the cobalt-catalyzed hydrogenation of lauronitrile, this has been observed [14]. The dependence of hydrogen concentration upon reaction performance may be complicated by many factors including reactor design and chemical characteristics of the catalyst, necessitating optimization for any given catalytic system. Indeed, three-phase reactions like nitrile hydrogenation in the liquid phase over heterogeneous catalysts represent a challenging area for reaction engineering, in general.

5.3.6.5 Conditions for secondary and tertiary amines

Often discussions of nitrile hydrogenations revolve around primary amine selectivity for commercial purposes. Selective preparation of secondary and tertiary amines can be valuable for a range of applications including surface active agents and disinfectants. Reaction conditions, starting with catalyst selection, can be tuned to favor secondary and tertiary amines. Noble metal catalysts tend to favor condensation reactions over base-metal catalysts [89]. Accordingly, in the hydrogenation of butyronitrile, Pd/C provided 97% selectivity to tributylamine, whereas Rh/C provided 100% selectivity to dibutylamine under comparable conditions, Table 5.3. Benzonitrile showed somewhat different reactivity with little tertiary amine formation; rather Pd and Rh exhibited complicated trends with an impact from solvent but still showed significant condensation reactions [89].

Table 5.3: Secondary and tertiary amine production [89].

Nitrile	Catalyst	Medium	*T* (°C)	H_2 press. (MPa)	Selectivity		
					Primary	Secondary	Tertiary
Butyronitrile	Rh/C	H_2O/NH_3	75–110	3.2–4.1	0	100	0
Butyronitrile	Pd/C	H_2O/NH_3	125	3.2–4.1	0	3	97
Butyronitrile	Pt/C	H_2O/NH_3	125	3.2–4.1	0	3	97
Benzonitrile	Rh/C	Octane	RT	0.34	0	100	0
Benzonitrile	Pd/C	Octane	RT	0.34	63	34	0
Benzonitrile	Pt/C	Octane	RT	0.34	0	94	0
Benzonitrile	Rh/C	H_2O	100–110	6.2–8.3	22	42	0
Benzonitrile	Pd/C	H_2O	100	2.1–3.4	19	49	0
Benzonitrile	Pt/C	H_2O	105	3.4–5.5	0	97	0

5.4 Catalysts

While heterogeneous nitrile hydrogenation catalysts have evolved significantly over the past several decades, the synthesis and selection of such catalysts have largely remained empirical. To effectively impact the rate, selectivity, and life of the catalysts, scientists and chemical practitioners, time and time again, have advantageously used base metals and/or precious metals as active species. Such catalysts have been deployed either in the supported or unsupported form. To further complicate things, structural promoters and activity and selectivity modifiers have been used to improve the efficacy of the reactions.

5.4.1 Slurry activated base metal catalysts

Activated base-metal catalysts, commonly known as Raney® (trademark of W.R. Grace) catalysts, are predominantly used in many high volume commercial nitrile hydrogenation reactions. This type of catalyst was first invented by Murray Raney [90], who sold the technology to W. R. Grace. The catalyst is prepared by first making an alloy of a base metal (e.g., nickel, cobalt, copper, and iron) with primarily aluminum (although silicon has also been used to prepare such alloys), at high temperature (1,200°C and above), in a fixed proportion to produce a eutectic and/or peritectic composition [91, 92]. Subsequently, the alloy is ground into fine particles and digested in a caustic solution to leach out aluminum in the form of sodium aluminate, leaving behind a very porous, skeletal, metallic structure. While the ratio of aluminum to base metal plays a significant role in the activity of the final catalyst composition, the effect of the concentration of caustic as well as the degree of dealumination, and the addition of promoters, may all influence the activity, selectivity, and life of the catalyst. A large volume of literature is available on the structure–activity relationship and stability of these catalysts, based on the widely varying preparation methods, and is beyond the scope of this article [93].

5.4.2 Fixed-bed activated base metal catalysts

Activated base metal catalysts, usually manufactured and commercially available in the slurry form, have an advantage over supported base-metal catalysts for being available in the pre-activated form, having high surface area, and relatively high activity at lower temperatures [94]. However, when it is necessary to obtain high selectivity and yield of an intermediate product (B) in a sequential reaction (A → B → C), it is desirable to run the reaction in a fixed-bed reactor [95]. As a result, attempts have been made to prepare fixed-bed activated base metal (e.g. Raney(R)) catalysts in different shape, size, and structure. Some of these attempts have resulted in fixed-bed Raney® catalysts, available commercially, one such offering being available from W. R. Grace in the form of pre-activated granules of various size fractions. The alloy, used in the fixed-bed Raney® catalyst, is prepared in the same way as it is done for the slurry catalyst. However, the alloy is then subjected to crushing, grinding, and sieving to obtain granules in the desired size range. Subsequently, the granules are activated to prepare a “core–shell” catalyst, where a thin shell, in the order of 0.6–1.2 mm thickness, is activated by leaching aluminum in a controlled fashion using caustic solution and leaving the unactivated core intact to ensure that the activated catalyst granules have adequate crush strength.

A more creative approach of making fixed-bed Raney® catalyst has been described by Cheng et al. [95, 96]. In this process, aluminum nickel alloy, in the form of powder, consisting of 58 wt% Al and 42 wt% Ni was mixed with high density polyethylene and mineral oil. The latter was added as a plasticizer for the polyethylene. The mixture was compounded at 150°C and extruded. The plasticizer was extracted using a suitable solvent and the extrudate was dried to increase its strength. The extrudate containing NiAl alloy in the polyethylene matrix was activated using 20% NaOH solution, as described by Freel et al. [92], providing the polymer-bound Ni catalyst with 50 $m^2.g^{-1}$ BET surface area. Such polyethylene-supported catalysts have been described elsewhere [97]. Unfortunately, such catalysts are only effective below the softening point of the polymer matrix, which is only 90°C for polyethylene. So instead of restricting the use of this catalyst to below the softening point of polyethylene, Cheng and coworkers calcined the catalyst in air at temperatures >900°C [95]. During this process, polyethylene was burned off and aluminum in the Ni catalyst was partially converted to α-alumina, which further increased the crush strength of catalyst. The degree to which the aluminum transformed to α-alumina depends on the calcination temperature and time. The calcined catalyst was activated in a 20 wt% NaOH solution at 90°C for 3 h and washed in DI water to maintain a pH <9 to provide a fixed-bed Raney(R) Ni catalyst that can be used at temperatures, greater than the softening temperature of polyethylene, i.e., 90°C.

O'Hare et al. proposed forming shaped catalysts to impart higher strength to relatively brittle activated base metal catalysts and at the same time provide heat sink for highly exothermic reactions. The base-metal–aluminum alloy in substantially the size and shape desired for the final massive, shaped catalyst. The alloy is then treated with an alkali solution to dissolve the aluminum portion of the alloy from a surface layer of the massive alloy [98]. This results in formation of a catalytically active layer of high surface area activated base metal catalyst on the surface of the massive alloy. Similar method of preparation has been described by Sauer et al. [48], wherein only a thin outer layer of 0.1–2.0 mm in thickness, of the base-metal–aluminum alloy, is activated to prepare the formed catalyst.

5.4.3 Hollow-activated metal catalyst

Hollow-activated base-metal catalyst is a relatively new approach for manufacturing fixed-bed activated base metal catalysts. One of the methods for the preparation of formed activated base metal catalyst, described in the literature, involves addition of the base metal, in powder form, as a binder to the base-metal–aluminum alloy to provide the necessary crush strength and attrition resistance of the formed catalyst [45].

The disadvantages of the formed activated base metal catalysts, described in the previous section, are that they have very high bulk densities and a significant portion of the catalysts, in the form of binder and/or base-metal–aluminum alloy, remain inert in the hydrogenation process. The addition of the binder, in the form of metal powder, limits the amounts of pore former (used to increase the surface area of the catalyst) that can be added without compromising the strength of the shaped catalyst. Furthermore, when a thin outer section of the formed catalyst is activated, the catalyst tends to lose activity rather rapidly due to the attrition of the outer surface of the catalyst. These and other factors including lower active surface area per unit volume of the catalyst and mass transfer limitations in three-phase reactions have tremendous economic disadvantage for commercial processes. To avoid these drawbacks, Ostgard et al. described a process for producing a hollow spherical form of the shaped activated base metal catalyst. In this process, an aqueous suspension of rapidly cooled base-metal–aluminum alloy, polyvinylalcohol, and glycerin was spray-coated on 4–5 mm polystyrene balls, fluidized in a flowing steam of air [99]. The alloy-coated polystyrene balls were dried in an air stream at 80°C and, subsequently, heated at 830°C for 1 h in a controlled nitrogen/air environment to burn off the polystyrene balls and sinter the alloy particles. The hollow spheres were then activated by standard activation procedure, described elsewhere, to produce active base metal catalyst.

The hollow type of activated base metal catalysts has the advantage of being cost-effective due to lower bulk density than the other types of fixed-bed activated base metal catalysts and high activity per unit weight of the catalyst. This type of catalyst has been beneficially used for a large number of hydrogenation, dehydrogenation, isomerization, and hydration reactions of organic compounds, including nitrile hydrogenation.

5.4.4 Unsupported iron and cobalt catalysts

Driven by reducing catalyst cost, DuPont, ICI, BASF, and Bayer have patented supported and unsupported iron and cobalt catalysts for large-scale commercial nitrile hydrogenation processes, including ADN hydrogenation to ACN and/or HMD [100–104]. Iron catalyst supported on pumice, reduced at 350°C, was used for producing primarily ACN and a small amount of HMD by the hydrogenation of ADN in the presence of toluene and liquid ammonia at 85°C [105].

Dewdney et al. disclosed a process for producing iron catalyst by fusing Swedish magnetite ore, containing 98.47 wt% iron oxide, at 1,590°C for 1 h [106]. The fused iron oxide product had the following composition:

Total iron content (%): 70.7%; Fe(II): 19.9, Fe(III): 50.8; iron oxide: 98.2, Al_2O_3: 0.2, SiO_2: 0.5, CaO: 0.1, V_2O_5: 0.2, atomic ratio of oxygen to iron: 1.36:1

The fused material was crushed to a specific sieve size and converted into active catalyst by first heating it at 350°C under nitrogen for 3 h and then reducing it at 450°C under flowing hydrogen at a flowrate of 10 $mL.min^{-1}.g^{-1}$ sample for 48 h. The reduced catalyst was cooled under nitrogen and used in the hydrogenation of ADN. Hydrogenation catalyst was also prepared using Labrador hematite. The inventors also described a few different variations of the reduction process including one in which hydrogen containing 1–3 vol% ammonia was used [106].

A similar process was described earlier by Kershaw et al. to prepare iron (containing about 3 wt% alumina) and cobalt catalysts [107]. The catalysts were used for ADN hydrogenation and the activity, selectivity, yield, and life were compared side by side. Higher yield of HMD was obtained with the iron catalyst compared to the cobalt catalyst (98.8% vs. 96.4%). The life of the iron catalyst was also longer than the cobalt catalyst measured as the throughput (51,000 g of ADN processed for Fe catalyst vs. 8,800 g for Co catalyst) at the same productivity (2.38 $kg.h^{-1}$ of diamine produced per liter of catalyst). Later on, Wu provided a more preferred form of the iron oxide catalyst precursor [104]. On a dry basis, the iron oxide catalyst precursor contained 93.4 wt% Fe_2O_3, 0.70 wt% FeO, 5.0 wt% SiO_2, and a total of 1.6% other oxides (Al_2O_3, CaO, MgO, MnO, Na_2O, K_2O, Li_2O, and TiO_2). The iron oxide was reduced at 460°C in 99 vol% hydrogen and 1 vol% ammonia for 22 h and subjected to ADN hydrogenation in a 1-L stirred autoclave containing 216 g, each, of ADN and ammonia. The reaction was carried out under hydrogen at an operating temperature and pressure of 150°C and

34.5 MPa (5,000 psig), respectively. A reaction rate of 3.3 g ADN.g catalyst^{-1}.h^{-1} was obtained [104].

5.4.5 Supported precious metal catalysts

By and large, industrial precious metal catalysts are supported, wherein the metal is dispersed on a high surface area metal oxide, mixed metal oxide, or carbon. One or more precious metals including Pt, Pd, Rh, Ir, and Ru have been expediently used as supported metal catalysts for the hydrogenation of nitriles. A plethora of supports have been used for nitrile hydrogenation; however, still the most prevalent industrial catalytic supports seem to be silica, alumina, and activated carbon. Supported precious metal catalysts are commonly prepared by a handful of preparation techniques including impregnation, coprecipitation, homogeneous deposition precipitation, and precipitation at constant pH [108]. For a detailed description of these techniques, the reader is referred elsewhere [109–111].

Depending on the type of reactor they are used in, heterogeneous precious metal hydrogenation catalysts are primarily of two kinds: fixed bed and slurry type. The former type of catalyst can be of different shapes, sizes, and forms, while the latter is always in the powder form. The production of such catalysts on the commercial scale with batch-to-batch replication of the activity, selectivity, thermal stability, and mechanical properties of the catalyst is extremely difficult. As a result, the catalyst vendors typically maintain the methods for the preparation of such catalysts as a trade secret, instead of disclosing such methods in open or patent literature.

Activity and selectivity of supported metal catalysts can be fine-tuned by changing the depth of penetration of active catalytic phase in formed catalysts. Based on the distribution of the active phase, the supported metal catalysts can be classified into four different categories, such as (a) eggshell, (b) egg yolk, (c) egg white, and (d) uniform. These metal distribution profiles can be conveniently generated by altering the pH on the surface of the catalyst [108]. Impregnation of the support is followed by drying, calcination, and reduction, and stabilization of the supported metal catalysts [112–115].

Only a handful of studies are available in the literature, which have reported the effect of catalyst configuration on the efficacy of nitrile hydrogenation reactions. By far, uniform and eggshell dispersion of the active metal(s) on a support are the most prevalent configurations of supported precious metal catalysts. In addition to preparative methods described above, eggshell catalysts are also synthesized by depositing "pre-prepared nanometal colloids" on supports. Such catalysts contain a thin layer (<250 nm) of metal particles on the surface of a suitable support and have been successfully used to reduce nitrile compounds [116].

It is well known in the literature that dibenzylamine, a secondary amine, is the primary product for the reduction of benzonitrile in the presence of heterogenous Pt catalysts [117]. Although the activity of Pt nanoparticles (~4 nm) dispersed on metal organic framework (MOF) MIL-101(Cr) support in eggshell configuration was lower than that of Pt/Al_2O_3 catalyst, Khajavi et al. observed that benzylamine, a primary amine, was selectively formed over dibenzylamine in the presence of this Pt-supported MOF catalyst system [117]. The change in selectivity for the Pt-MIL-101(Cr) catalyst has been attributed to steric hindrance of the transition state adsorbed on the Pt nanoparticles.

While selecting catalysts for nitrile hydrogenation in real-world industrial applications, selection of active species, support, promoters, and catalyst preparation techniques and conditions is deemed extremely important for the cost-effectiveness of the process and must be approached systematically and judiciously to enable high activity, selectivity, stability, and life of the catalyst.

5.4.6 Homogeneous catalysts

Beyond heterogeneous catalysts, numerous homogeneous catalysts have been developed for nitrile hydrogenation; indeed, this has been a very active field of research in the past 10 years and has recently been reviewed [5]. In 1969, Dewhirst provided one patent example of a ruthenium phosphine complex catalyzing the hydrogenation of benzonitrile in 29% isolated yield [118]. Grey et al. reported seminal work, in 1981, on the hydrogenation of nitriles by a homogeneous Ru phosphine hydride complex, invoking a hydride-transfer mechanism [119]. The work demonstrated the influence of 18-crown-6 on improving conversion and yield in batch reactions and was conducted under fairly mild conditions at 90°C, 0.72 MPa, 20 h, using ≤0.3 mol% catalyst. Ruthenium has been a notable metal in the literature of homogeneous catalysts for nitrile hydrogenation. Beatty and Paciello defined nitrile reduction chemistries catalyzed by Ru-hydride phosphine complexes of the form $RuH_2L_2(PR_3)_2$ ($L = H_2$ or PR_3) [120–125] in the 1990s. Interest was revitalized in these complexes in the 2010s with a report of modified ligands and mechanistic insight [126]. Beller explored readily adoptable Ru catalyst systems employing a variety of ligands for Ru catalyzed nitrile hydrogenation at 8.0 MPa H_2, 80°C, and 0.1 mol% catalyst, wherein 1,1′-bis(diphenylphosphino)ferrocene was deemed the optimal ligand and the reactions incorporated 10 mol% of a strong base (KO^tBu) [127]. The authors commented that the reactions did not require ammonia gas, which is typical of heterogeneous catalyst systems, though notable exceptions exist (*vide supra*). Subsequently, Ru pincer complexes were studied as nonclassical hydride complexes competent in nitrile hydrogenation reactions [128]. In 2012, Milstein reported a Ru pincer complex for nitrile hydrogenation coupled with amine attack to selectively form imines [77] and proposed ligand-cooperativity and heterolytic activation of hydrogen.

Significant effort has been devoted in recent years to the design of catalytic nitrile hydrogenation employing base-metal homogeneous catalysts. The utilization of base

Table 5.4: Hydrogenation of nitriles to primary amines with a homogeneous Fe catalyst [80].

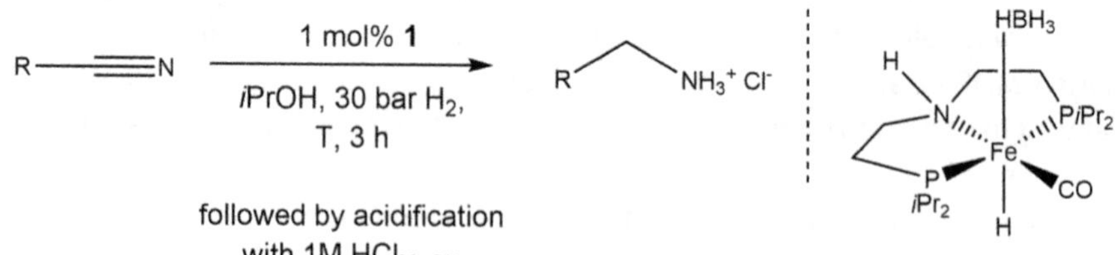

R–C≡N → (1 mol% **1**; *i*PrOH, 30 bar H_2, T, 3 h; followed by acidification with 1M HCl_{MeOH}) → $RCH_2NH_3^+ Cl^-$

Nitrile	Primary Amine	T (°C)	Conv. (%)	Yield (%)
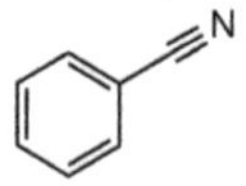 benzonitrile	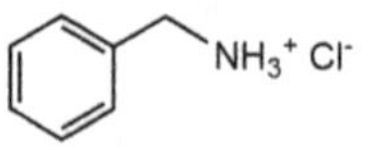 benzylamine·HCl ($NH_3^+ Cl^-$)	70	>99	97
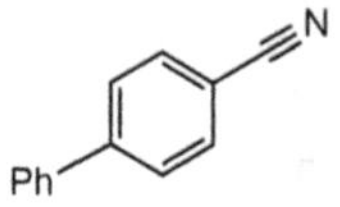 4-Ph-benzonitrile	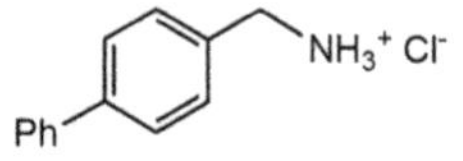 4-Ph-benzylamine·HCl ($NH_3^+ Cl^-$)	70	>99	99
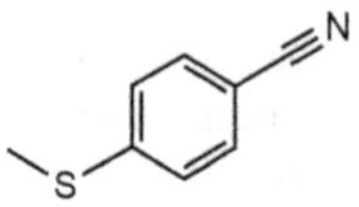 4-MeS-benzonitrile	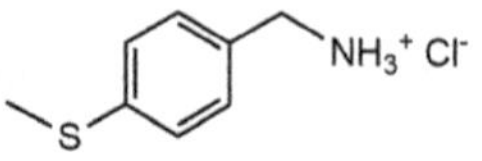 4-MeS-benzylamine·HCl ($NH_3^+ Cl^-$)	100	>99	81
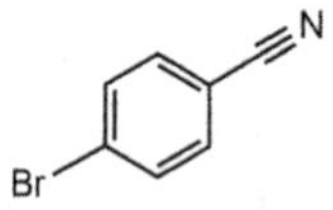 4-Br-benzonitrile	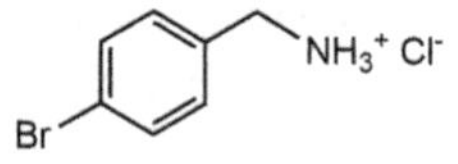 4-Br-benzylamine·HCl ($NH_3^+ Cl^-$)	100	>99	88
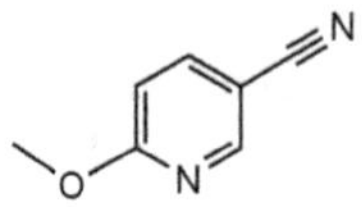 6-methoxy-3-cyanopyridine	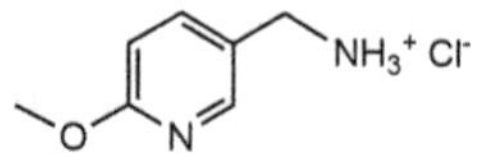 (6-methoxypyridin-3-yl)methylamine·HCl ($NH_3^+ Cl^-$)	100	>99	92
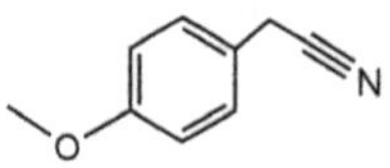 4-MeO-phenylacetonitrile	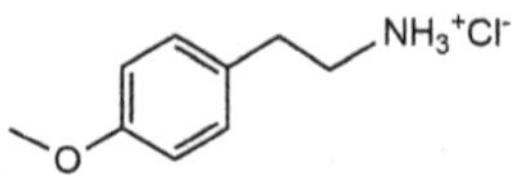 2-(4-MeO-phenyl)ethylamine·HCl ($NH_3^+Cl^-$)	70	>99	89
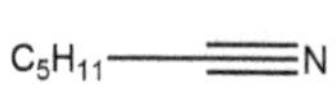 C_5H_{11}–C≡N	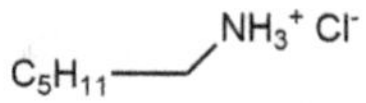 $C_5H_{11}CH_2NH_3^+ Cl^-$	70	>99	85

metals offers potential economic benefit in light of their greater abundance and lower cost compared to precious metals. Furthermore, in the context of pharmaceutical manufacturing, base-metal catalysts, particularly iron, copper, and manganese, offer an advantage in that regulatory agencies stipulate extremely low residual metal content for precious metals but have a greater tolerance for iron, copper, and manganese,

Table 5.5: Hydrogenation of nitriles to secondary imines with a homogeneous Fe catalyst [132].

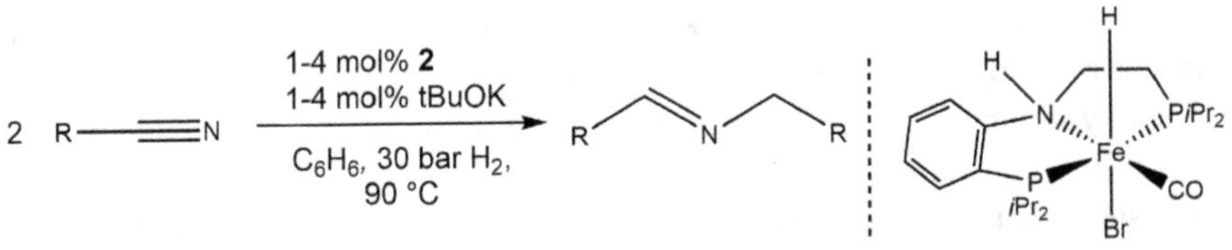

2 R—≡N → (1-4 mol% **2**, 1-4 mol% tBuOK; C_6H_6, 30 bar H_2, 90 °C) R⁀N⁀R

2

Nitrile	Product	Time (h)	Conv. (%)	Yield (%)
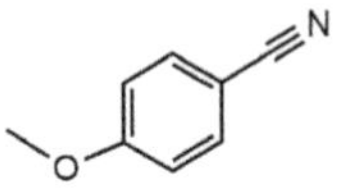	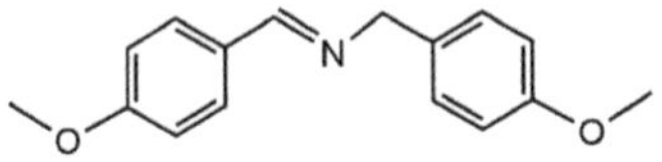	<36	>99	97
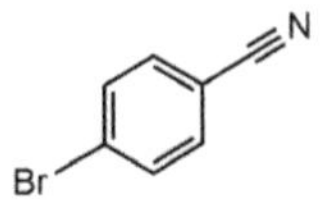	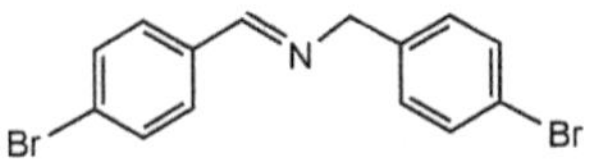	12	>99	99
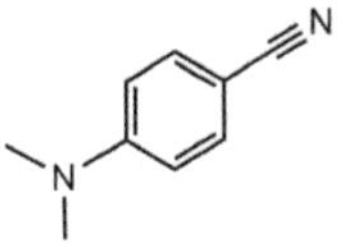	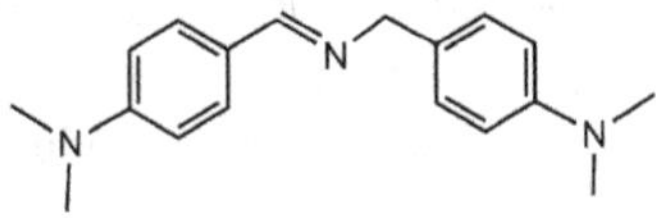	18	81	52
 C_3H_7—≡N	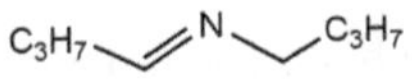 C_3H_7⁀N⁀C_3H_7	36	32	32

which are deemed less toxic [129]. Beller made the first report of a homogeneous Fe catalyst for selective nitrile hydrogenation, employing a ligand to facilitate heterolytic activation of hydrogen (Table 5.4) [80]. The reactions proceeded at 3 MPa H_2 and 70–100°C in isopropanol with no addition of strong base and are appealing for scale-up.

Nearly simultaneous to the Beller report, Chakraborty and Berke described Mo and W catalysts for reduction of nitriles to secondary imines and other products [130]. Interestingly, they used a ligand, which could potentially promote a cooperative outer-sphere activation mechanism similar to the Beller and Milstein systems but did not delve into the mechanism in their initial report. In addition to the above, Milstein communicated Co pincer complexes [82] and an iron pincer complex for nitrile hydrogenation [131], each employing catalytic quantities of hydride reagents and base to initiate and enhance the reaction. Additionally, each system incorporated ligands with the potential for cooperativity in the heterolytic activation of H_2 as described in the mechanism section above. A recent publication from Milstein described an iron catalyst for selective imine formation (Table 5.5) [132]. The catalyst was structurally and electronically similar to the Beller system but differentiated in its incorporation of a phenyl group in the ligand backbone. This difference along with use of an alternate solvent led to a stark change in reaction selectivity. (Note, while the Br ligand also differentiates this complex, it is released on treatment with base as part of catalyst activation/initiation.)

Functional group tolerance is a key attribute in development of a catalytic system and was good to exceptional with the homogeneous catalysts developed, to date. It is especially notable that Ar–Br was not significantly cleaved under reducing conditions in the presence of the transition metal catalysts. By contrast, heterogeneous catalysts such as Raney® Ni, Pd/C, and Pt/C are known to suffer from incompatibility with aryl halides, marking an advantage for homogeneous catalysts reported for nitrile hydrogenation.

5.5 Reactor design

Selection and design of chemical reactors is at the forefront of new chemical reaction technologies. Even the best possible catalyst can perform suboptimally, if the selection of reactor is not correct or it is subjected to reaction conditions, which cause rapid deactivation of the catalyst. Many physical, chemical, and economic factors influence reactor selection. Nitrile hydrogenation has been conducted in a variety of reactor types [63]:

1. *Batch Reactors*
   Shaker tubes, stirred batch, fed-batch, and semi-batch reactors
2. *Continuous Stirred Tank Reactors (CSTR)*
   Mechanically agitated and slurry bubble column reactors (SBCR)
3. *Plug flow reactors*
   Trickle bed and packed bubble column reactors
4. *Fluidized bed reactors*
   Circulating fluidized bed and fluid bed reactors

5.5.1 Slurry bubble column reactors

Hydrogenation of ADN with Raney® Ni catalysts for nylon production is conducted in SBCR [133]. Bubble column or transport reactors are typically CSTRs, where catalyst particles are slurried or suspended in a liquid medium (substrate or reactant), while the gas is bubbled through the column. The efficacy of the process depends on good mixing and diffusion of the gas molecules from the gas phase, through the liquid phase and on to the surface of the particulate catalyst, through the liquid phase. The reaction takes place on the surface of the heterogeneous catalyst, where the liquid and the gaseous moieties adsorb and come in contact with one another (Figure 5.17). The particulate catalyst is suspended in the liquid, while the gas and liquid streams

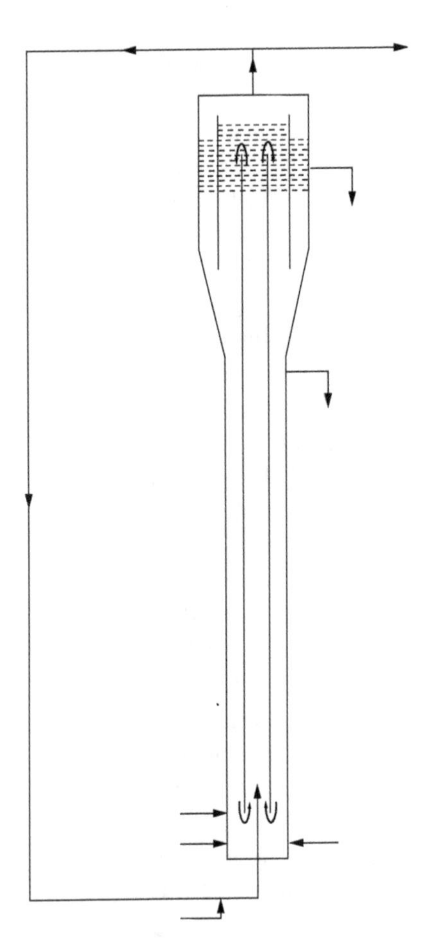

Figure 5.17: Slurry bubble column reactor used in the production of HMD [134].

traverse in the upward direction. SBCR, typically, comprise a well-designed draft tube in the center of the reactor to provide good mixing of the gas, liquid, and solid. The reactor is flared at the top to disengage the gas from the liquid and enable internal recirculation of the liquid by recycling the latter to the bottom of the draft tube. Another advantage of an SBCR is constant replenishment of the catalyst and removal of an equivalent amount of catalyst, thereby removing a portion of the deactivated catalyst and increasing the uptime of the process.

When a SBCR is used for the hydrogenation of ADN, commercially it is performed in the liquid phase in the presence of an activated base metal catalyst (e.g. Raney(R) nickel) and an aqueous alkali metal hydroxide solution, such as NaOH. ADN, hydrogen, and aqueous caustic soda solution are continuously fed to the reactor comprising HMD and Raney® nickel catalyst. The reaction is carried out under elevated temperature (60–100° C) and pressure (2–5 MPa). Continuous addition of aqueous NaOH solution is necessary to suppress the formation of secondary and tertiary amines and extend the life of the catalyst. It is continuously added to the reactor and maintained in the range of

0.2–12 mol.kg^{-1} catalyst, while the concentration of water is maintained in the range of 2–130 mol.mol^{-1} of the base [9, 134]. By contrast, Raney® Co is more resistant to deactivation [135] and may be used in a CSTR without continuous addition of inorganic base, in one case showing negligible deactivation for >1,800 g feed.g of catalyst^{-1} [136].

Designing of an SBCR is immensely complicated due to the fact that phase holdup, flow regime, bubble size distribution, coalescence characteristics, gas–liquid interfacial area, interfacial mass transfer coefficients, heat transfer coefficients, and dispersion coefficients influence the design of an SBCR [137, 138]. A thorough investigation was undertaken by CNRS/Rhone-Poulenc group in order to study the effect of the above parameters on the catalytic hydrogenation of ADN in a continuous SBCR fitted with an internal draft tube [139]. Systematically, they carried out an extensive experimental program to address the reaction kinetics, hydrodynamics, mass transfer, and design, control, and operation of this three-phase reaction system. While the reactor modeling was done in a continuous bubble column reactor, the hydrodynamics was performed in a transparent PVC tube in a batch-liquid system under atmospheric temperature and pressure. Nitrogen instead of hydrogen was used in the gas phase, while water and reaction solvent were used as the liquid phase and Raney(R) nickel catalyst as the solid phase to determine the gas holdup in the presence and absence of a draft tube, and as a function of solid loading, and the swarm velocity of the bubbles. The authors claimed that the hydrodynamics of a pressurized bubble column reactor can be obtained using an atmospheric system as the liquid flowrate was low; there were others who contradicted that claim before, for example, Wilkinson and Dierendonck's extensive study on the effect of pressure and gas density on gas holdup in bubble columns [140]. They argued that when high-pressure industrial bubble columns are designed based on data obtained under atmospheric pressure, it causes an underestimation of gas holdup and mass transfer. Perhaps that's why the CNRS/Rhone-Poulenc group found that no correlation accurately predicted the gas holdup in the reaction solvent [139].

Like hydrodynamic models and mass transfer correlations, intrinsic kinetic model development is equally important for designing reactors, be it batch, plug flow, or continuous reactors [141, 142]. Batch and semicontinuous reactor configurations have been used to derive an intrinsic kinetic reaction model for three-phase hydrogenation of ADN in the presence of Raney® Ni catalyst [141, 142]. A Langmuir–Hinshelwood kinetic model was selected to delineate the two-step hydrogenation process (ADN → ACN → HMD). Different rate-limiting steps including adsorption, reaction, desorption, dissociative adsorption of atomic hydrogen or molecular adsorption of dihydrogen and single or dual sites for nitrile and hydrogen adsorption were considered, while modeling the reaction kinetics. Nonlinear optimization using Marquardt algorithm showed that the experimental data fit the kinetic model based on surface reaction between dissociated hydrogen and nitriles adsorbed on different sites being the rate-determining step. However, the model discrimination based on standard deviation did not adequately support one over the other rate-controlling mechanism and could likely be due to inadequacy of Langmuir models in liquid phase reactions.

5.5.2 Packed bed reactors

While slurry bubble column and stirred tank reactors are preferred for ADN hydrogenation process when activated base metal catalysts are used, high-pressure fixed or packed bed reactors are employed in the Fe-catalyzed process. ADN has been hydrogenated in packed bed reactors in the presence of excess of hydrogen and anhydrous ammonia, at temperatures ranging from 85 to 185°C and pressures of 27.6–41.4 MPa (4,000–6,000 psig) (Figure 5.18) [104]. Excess ammonia was used as a solvent to remove heat generated in this exothermic reaction and suppress the formation of undesirable by-products, which are difficult to remove (*vide supra*). The iron oxide catalyst precursor, used in this process, was pre-reduced using 99% hydrogen and 1% ammonia by volume at 460°C for 22 h. The completion of the reduction process was judged by the amount of water collected from the off-gas [104].

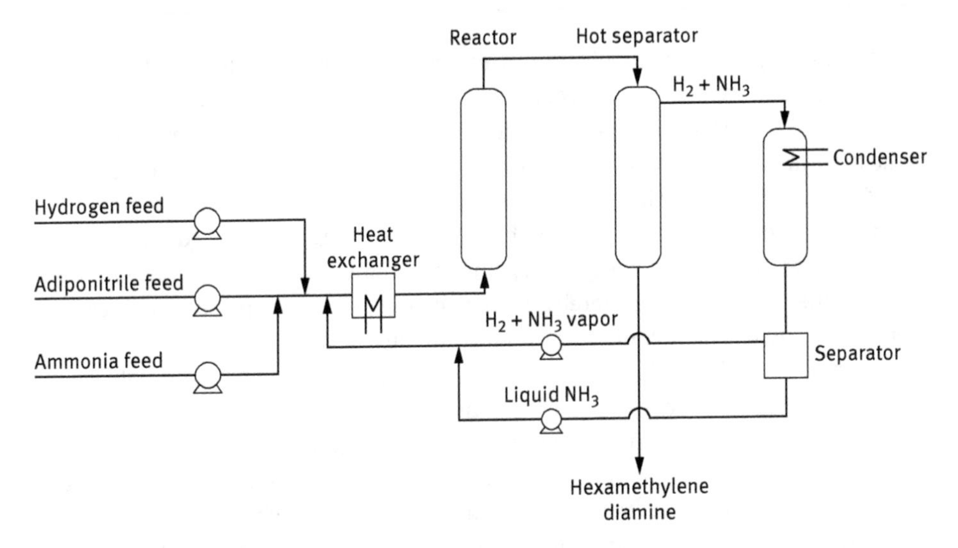

Figure 5.18: Continuous fixed-bed reduction of adiponitrile to 1,6-hexanediamine [143].

Traditionally, specialty and fine chemical industries use batch three-phase slurry reactors for hydrogenation processes. However, with the quest to improve productivity, use green chemistry, and improve process economics, even these traditional industries are breaking away from the conventional processes and taking on challenging tasks of addressing reactor design and reaction engineering, process intensification, hydrogen mass transfer, removing solvents, or changing reagent addition strategies in addition to finding novel catalysts or developing new reaction pathways. This is even true for nitrile hydrogenation processes, wherein specialty amine-manufacturing companies are making small changes by replacing batch with semi-batch processes by using Buss-Loop reactors (Table 5.6) [144].

Table 5.6: Comparison of process approaches to increase productivity.

	Lab2 L	Original Plant Process	Buss Loop reactor 300 L
Process class	**Batch**	**Semi-batch**	**Semi-batch**
Temperature	100°C	140°C	140°C
Pressure	35 barg	50 barg	50 barg
kla (s^{-1})	unknown	0.40	1.0
Net reaction time (h)	2.2	1.0	1.0
Maximum heat duty on reactor ($kW.kg^{-1}$)	0.70*	0.50	0.50
Productivity (kg product.kg $catalyst^{-1}.h^{-1}$)	30.0	50.0	50.0
Primary amine	92.0%	99.17%	99.32%
Nitrile	0.50%	0.05%	0.05%
Secondary amine	2.39%	0.05%	0.16%
Other by-products	5.11%	0.73%	0.47%

More and more, process intensification is also being explored in academia to perform one-pot reaction of nitrile hydrogenation and separation of specialty amines. Xie et al. reported the use of gas-expanded liquids (e.g., CO_2–ethanol system) as tunable media for primary aromatic amine formation and separation [145]. They have demonstrated the potential for using modest pressures of CO_2 to facilitate hydrogenation of benzonitrile and phenylacetonitrile with $NiCl_2/NaBH_4$ in CO_2-expanded ethanol system as well as to separate the primary amines from the reaction media. CO_2 also protected the amine moieties thereby, increasing the yield of the primary amines and suppressing the production of the secondary amines.

Much progress has been made in the chemical, petrochemical, and petroleum industries, so far as novel reactor design, new reaction engineering concepts, and economical process intensifications are concerned. Specialty chemical, pharmaceutical, and agrichemical industries are still lagging behind in these areas; however, it will probably take one researcher or one scientist to change one company, one process, one reactor at a time. Change is inevitable but it is just a matter of time.

5.6 Outlook on trends and needs in nitrile hydrogenation

The field of nitrile hydrogenation is significantly influenced from two very different perspectives: (1) the cost-effective, commercial manufacture of amine products at scales from tons to megatons (e.g., HMD for Nylon 6,6) and (2) practical laboratory and small-volume specialty chemical production. These disparate perspectives suggest diverse needs for future development.

5.6.1 Catalytic process technologies

As stated above, two primary catalyst systems are used for large-scale production of HMD via ADN hydrogenation [3]. In one case, activated base-metal catalysts, like Raney® Ni, are utilized in medium pressure (5–10 MPa) reactors at moderate temperatures (e.g. < 100°C). These reactions avoid the use of ammonia to suppress secondary amine formation by careful optimization of solvent and additives, e.g. alcohols and aqueous sodium hydroxide. A challenge in these applications from a green chemistry point of view has been handling the strong base additive waste. Such caustic materials are a challenge for incinerators / heat recovery units with regard to materials of construction and also for handling of incinerator by-products. While significant effort has gone into the optimization of these commercial processes, even greater value could be derived from environmentally-friendly alternatives to alkali metal hydroxide and / or new processing technologies for recovery of the alkali components in a cost-effective manner. Diverting strong base away from waste streams represents an opportunity for enhancing the sustainability of commercial processes. In an alternate commercial process for hexamethylenediamine production, iron or cobalt based unsupported catalysts have been utilized in high temperature (even > 200°C), high-pressure reactors with a large excess of ammonia. Ammonia is utilized to limit secondary amine formation and also as a medium for heat removal. An advantage of processes using iron-based catalysts is low cost associated with the active metal compared to the higher cost of nickel-based catalysts but the use of ammonia necessitates higher pressure reactors with higher capital investment and limits the volumetric efficiency of the reactors. In analogy to activated base-metal catalysts like Raney® Ni, an opportunity exists, if operating conditions or additives could be developed that would allow removal of ammonia from the process, while still maintaining or improving yield and productivity. In such a case, lower pressure reactors could be utilized and still increase in hydrogen partial pressure could be attained. It is notable that homogeneous iron catalysts are under development and show promising yields of primary amines from aliphatic nitriles, *in the absence of ammonia*. There have been limited discussions in the literature on the use of such catalysts in continuous reactors and their robustness toward deactivation and such investigations are a critical prerequisite for serious consideration in a commercial setting. In a related direction of research, it would be valuable if the intrinsic activity of heterogeneous iron-based catalysts could be improved, as the high-temperature operating conditions are likely utilized because of low activity [3]. Multiple research directions could be pursued for enhancing activity, including increase of active surface area to increase the volumetric activity and also modification of the chemical composition (use of promoters) of the iron-based catalysts to increase intrinsic activity. Optimization of commercial processes is significantly dependent upon catalyst lifetime studies and mechanistic studies to understand and inhibit deactivation modes are warranted.

Similarly, mechanistic studies may guide future catalyst design/invention efforts. For example, it has been suggested in the literature that enhancement of primary amine selectivity over Ni occurs when the ensemble size is reduced by "poisoning" of the catalyst surface [40, 68]. New opportunities exist for the rational design of catalyst structures and compositions to optimize primary amine selectivity. If small metal ensembles improve selectivity, it follows that resistance to metal particle sintering could be an important aspect of catalyst design. The thermal stability of catalysts under conditions of operation, along with resistance to poisoning, facility of regeneration, and feasibility of commercial manufacturing are all important attributes of catalyst design for future research efforts.

Acknowledging that the catalyst is just one component of a successful reaction system, the development of coupled reaction/separation schemes is a promising area of development. For example, biphasic liquid reaction systems may lead to enhanced selectivity of desired products. In one case, it has already been demonstrated that benzonitrile reduction coupled with hydrolysis in a biphasic liquid system gave high selectivity to benzyl alcohols [146]. A particular opportunity exists in coupling reaction and separation in the previously mentioned scheme for the selective hydrogenation of just one end of the ADN to form ACN, a potential intermediate to Nylon 6.

5.6.2 Nitrile hydrogenation for synthesis

The demands of the synthetic chemist or for the production of small volume products for high-value applications like pharmaceuticals, electronic chemicals, and agrichemicals are significantly different from commodity chemical production scenarios. Here, practitioners seek reaction parameters that are easily implemented in multipurpose facilities (or labs) with low-cost equipment. The use of anhydrous ammonia and/or pyrophoric solid catalysts like activated base metal catalysts engenders significant risk for those lacking expertise. Another inhibiting factor in these scenarios is access to pressure reactors, especially high-pressure reactors above 0.6 MPa. Indeed, it can be argued that the continued use of stoichiometric aluminum hydride reagents for the reduction of nitriles is connected to their utility in atmospheric pressure reactors. An active area of research in nitrile hydrogenation is the development of homogeneous catalysts, which address some of these barriers. The homogeneous catalysts can be manipulated by standard organic-laboratory practices, in contrast to Raney® Ni, and in many cases offer good selectivity to desired products without the use of anhydrous ammonia. However, in most cases described above, homogeneous catalysts have low activity when hydrogen pressures are below 1 MPa and, while many studies utilized pressures above 3 MPa, they still required long reaction times. Future research will surely address the shortcomings of the existing homogeneous catalyst systems. It is unclear from most of the recent homogeneous catalyst literature for nitrile hydrogenation, if long reaction times are associated with slow intrinsic kinetics or with catalyst deactivation. Accordingly, additional research should be conducted to better characterize catalyst fate as a function of reaction progress or continuous reactor operation. It is encouraging that in some cases homogeneous catalysts have been used at 4,000 h^{-1} [127].

In the context of fine chemical or biologically active molecule synthesis, functional group tolerance is of significant concern. Homogeneous catalysts have already demonstrated desirable functional group tolerance but further studies to define the full scope of functional group tolerance are warranted. Another concern with the use of homogeneous catalysts, particularly in pharmaceutical synthesis applications, is the potential for residual metal content. Here, the use of nontoxic, base-metal catalysts, e.g., Fe and Mn, could have significant value. The regulatory concern for such metals is much lower than with nitrile hydrogenation catalysts like Ni, Pd, or Pt, which must be limited to very low ppm levels. Further development of nontoxic base-metal catalysts for ease of manipulation and to enhance reactivity at low pressure operation would certainly be desirable.

Nitrile hydrogenation serves as a highly atom-efficient transformation for the synthesis of amines but clearly relies on availability of nitrile molecules. Accordingly, future developments in nitrile synthesis would yield additional value for nitrile hydrogenation. Numerous methods exist for the synthesis of nitrile molecules including amide dehydration, aldehyde plus hydroxylamine, cyanation of aryl and alkyl halides [147], hydrocyanation of olefins [148], and, more recently, the cross-metathesis of acrylonitrile and olefins [149] and a unique application of rhenium nitrides [150], but challenges remain with regard to functional group tolerance and the generality of the synthetic options. Developing safe and practical synthetic methods for nitrile molecules will reap additional benefit as nitrile hydrogenation technologies continue to advance.

References

[1] Weissermel K, Arpe H-J. Industrial Organic Chemistry, 2nd edn. Weinheim, Wiley VCH, 1993.

[2] Sabatier P, Senderens JB. Application to nitriles of the method of direct hydrogenation by catalysis; synthesis of primary, secondary, and tertiary amines. Compt Rend 1905, 140, 482–6.

[3] De Bellefon C, Fouilloux P. Homogeneous and heterogeneous hydrogenation of nitriles in a liquid phase: Chemical, mechanistic, and catalytic aspects. Catal Rev Sci Eng 1994, 36, 459–506.

[4] Krupka J, Pasek J. Nitrile hydrogenation on solid catalysts – New insights into the reaction mechanism. Curr Org Chem 2012, 16, 988–1004.

[5] Bagal DB, Bhanage BM. Recent advances in transition metal-catalyzed hydrogenation of nitriles. Adv Synth Catal 2015, 357, 883–900.

[6] Sriram P, Hyde B, Merrill B, Woo M. Chemical Economics Handbook: Hexamethylene-Adiponitrile. London, IHS Chemical; 2017.

[7] Bivens D, Patton L, Thomas W, inventors. E. I. du Pont de Nemours and Co., assignee. Hydrogenation of adiponitrile. US patent US3758584 1973.

[8] Brake LD, inventor. Hydrogenation of adiponitrile over alkali-modified cobalt catalyst. US patent 3773832. 1973.

[9] Cutchens CE, Lanier LH, inventors. Monsanto, assignee. Catalyst separation in production of amines. US patent US4429159 1984.

[10] Cutchens CE, Mathews MJ, Sowell MS, inventors. Monsanto, assignee. Production and separation of amines. US patent US4491673 1985.

[11] Herzog BD, Smiley RA. Hexamethylenediamine. Ullmann's Encyclopedia of Industrial Chemistry, Wiley-VCH Verlag GmbH & Co. KGaA, 2000.

[12] Allgeier AM, Duch MW. Reactivity and surface analysis studies on the deactivation of Raney Ni during adiponitrile hydrogenation. Chem Ind (Dekker) 2001, 82, 229–39.

[13] Orchard JP, Tomsett AD, Wainwright MS, Young DJ. Preparation and properties of Raney nickel-cobalt catalysts. J Catal 1983, 84, 189–99.

[14] Volf J, Pasek J. Hydrogenation of nitriles. Stud Surf Sci Catal 1986, 27, 105–44.

[15] Lin Y-J, Schmidt SR, Abhari R, inventors. W. R. Grace & Co.-Conn., assignee. Synthesis of non-cyclic aliphatic polyamines. US patent 5105015. 1991.

[16] Koch TA, Krause KR, Sengupta SK, inventors. E. I. du Pont de Nemours & Co., USA. assignee. A process and cobalt catalyst for the continuous hydrogenation of adiponitrile to hexamethylenediamine and optionally to aminocapronitrile patent WO9843940A1. 1998.

[17] Merrill B, Woo M, Khan Y, Sriram P. Chemical Economics Handbook: Caprolactam. London, IHS Chemical; 2017.

[18] McCoy M. Slowly Changing How Nylon is Made. Chemical and Engineering News 2000, 78, 32–4.

[19] Rigby GW, inventor. E. I. du Pont de Nemours & Co., assignee. Aliphatic Amino-ntriles and Process of Producing Them. US patent US2208598. 1940.

[20] Rigby GW, inventor. E. I. du Pont de Nemours & Co., assignee. Preparation of Omega-Amino Nitriles. US patent US2257814. 1941.

[21] Allgeier AM, inventor. Invista North America S.A R.L., USA. assignee. Use of modifiers in a dinitrile hydrogenation process patent US20050101797A1. 2005.

[22] Jia Z, Zhen B, Han M, Wang C. Liquid phase hydrogenation of adiponitrile over directly reduced Ni/SiO2 catalyst. Catal Comm 2016, 73, 80–3.

[23] Ziemecki SB, inventor. E. I. du Pont de Nemours & Co, assignee. Selective low pressure hydrogenation of a dinitrile to an amino nitrile. US patent 5151543. 1992.

[24] Ionkin AS, Ziemecki SB, Harper MJ, Koch TA, inventors. E. I. du Pont de Nemours & Co., USA. assignee. Hydrogenation process and catalysts for the manufacture of aminonitriles from dinitriles patent WO9947492A1. 1999.

[25] Ionkin AS, inventor. E. I. du Pont de Nemours & Co., USA. assignee. Aminonitrile production from partial hydrogenation of dinitrile and improving yield and selectivity to aminonitrile patent US6455724B1. 2002.

[26] Ionkin AS, inventor. E. I. du Pont de Nemours & Co., assignee. Aminonitrile Production US patent US6506927. 2003.

[27] Ionkin AS, inventor. E.I. du Pont de Nemours and Co., USA. assignee. Partial hydrogenation process and catalysts for the manufacture of aminonitriles from dinitriles patent US20030065209A1. 2003.

[28] Ionkin AS, inventor. E. I. du Pont de Nemours & Co., USA. assignee. Partial hydrogenation process and catalyst system for the manufacture of aminonitriles from dinitriles patent US6506927B1. 2003.

[29] Verhaak MJFM, van Dillen AJ, Geus JW. The selective hydrogenation of acetonitrile on supported nickel catalysts. Catal Let 1994, 26, 37–53.

[30] Greenfield H. Catalytic Hydrogenation of Butyronitrile. I&EC Product Research and Development 1967, 6, 142–4.

[31] Janshekar H, Greiner E, Kumamoto T, Zhang E. Chemical Economics Handbook: Alkylamines: IHS Chemical; 2014.

[32] Segobia DJ, Trasarti AF, Apesteguia CR. Conversion of butyronitrile to butylamines on noble metals: Effect of the solvent on catalyst activity and selectivity. Catal Sci Technol 2014, 4, 4075–83.
[33] Vedage GA, Armor JN, inventors. Hydrogenation of nitriles to tertiary amines. US patent 5672762. 1997.
[34] Segobia DJ, Trasarti AF, Apesteguia CR. Synthesis of n-butylamine from butyronitrile on Ni/SiO2: Effect of solvent. J Braz Chem Soc 2014, 25, 2272–9.
[35] Thomas-Pryor SN, Manz TA, Liu Z, Koch TA, Sengupta SK, Delgass WN. Selective hydrogenation of butyronitrile over promoted Raney nickel catalysts. Chem Ind (Dekker) 1998, 75, 195–206.
[36] Global Vitamin B1 (Thiamine) Market Research 2011–2022: Explore Market Research; 2017 5/26/2017.
[37] Eggersdorfer M, Adam G, John M, et al. Vitamins. Ullmann's Encyclopedia of Industrial Chemistry, Wiley-VCH Verlag GmbH & Co. KGaA, 2000.
[38] Bonrath W, Medlock J, Schütz J, Wüstenberg B, Netscher T. Hydrogenation in the vitamins and fine chemicals industry – An overview. In: Karamé I, ed. Hydrogenation: InTech; 2012.
[39] Degischer OG, Roessler F, inventors. Roche Vitamins, assignee. Modification of a hydrogenation catalyst patent US6521564. 2003.
[40] Ostgard DJ, Roessler F, Karge R, Tacke T. The treatment of activated nickel catalysts for the selective hydrogenation of pynitrile. Chem Ind (Boca Raton, FL, U S) 2007, 115, 227–34.
[41] Gillet JP, Kervennal J, Pralus M. New process for isophoronediamine synthesis. In: Guisnet M, Barbier J, Barrault J, et al., eds. Studies in Surface Science and Catalysis. New York, Elsevier, 1993, 321–8.
[42] Ostgard D, Berweiler M, Roder S, et al., inventors. Degussa, assignee. Process for the preparation of 3-aminomethyl-3,5,5-trimethylcyclohexylamine. US patent US6437186. 2002.
[43] Huthmacher K, Schmitt H, inventors. Degussa A.G., assignee. Method of preparing 1,3,3-trimethyl-5-oxo-cyclohexane carbonitrile. US patent US5091554. 1992.
[44] Herkes FE, Kourtakis K, inventors. E. I. Du Pont De Nemours and Company, assignee. Preparation of isophorone diamine. US patent US5491264. 1996.
[45] Haas T, Burmeister R, Arntz D, Weber KL, Berweiler M, inventors. Degussa, assignee. Process for the production of 3-aminoethyl-3,5,5-trimethylcyclohexyl amine. US patent US5679860. 1997.
[46] Disteldorf J, Hubel W, Broschinski L, inventors. Process for the preparation of primary mono- and diamines from oxo compounds. US patent US4429157. 1984.
[47] Haas T, Weber KL, Stadtmuller K, Hofen W, Vanheertum R, inventors. Degussa-Huls AG, assignee. Process for the production of amines from imines of nitriles. US patent US6011179. 2000.
[48] Sauer J, Haas T, Keller B, et al., inventors. Degussa A.G., assignee. Shaped metal fixed-bed catalyst, a process for its preparation and its use. US patent US6337300. 2002.
[49] Haas T, Arntz D, Most D, inventors. Degussa Aktiengesellschaft, assignee. Process for the preparation of isophoronediamine. US patent US5504254. 1996.
[50] Leuckart R. Ueber eine neue Bildungsweise von Tribenzylamin. Berichte der deutschen chemischen Gesellschaft 1885, 18, 2341–4.
[51] Heuer L. Benzylamine. Ullmann's Encyclopedia of Industrial Chemistry. Wiley-VCH Verlag GmbH & Co. KGaA, 2000.
[52] Hegedűs L, Máthé T. Selective heterogeneous catalytic hydrogenation of nitriles to primary amines in liquid phase: Part I. Hydrogenation of benzonitrile over palladium. Appl Catal A 2005, 296, 209–15.
[53] Saito Y, Ishitani H, Ueno M, Kobayashi S. Selective hydrogenation of nitriles to primary amines catalyzed by a polysilane/SiO2-supported palladium catalyst under continuous-flow conditions. Chem Open 2017, 6, 211–5.
[54] Li Z, Fang L, Wang J, Dong L, Guo Y, Xie Y. An improved and practical synthesis of tranexamic acid. Org Proc Res Dev 2015, 19, 444–8.

[55] Sera M, Yamashita M, Ono Y, et al. Development of large-scale synthesis using a palladium-catalyzed cross-coupling reaction for an isoquinolone derivative as a Potent DPP-4 Inhibitor. Org Proc Res Dev 2014, 18, 446–53.
[56] Saravanan M, Satyanarayana B, Reddy PP. An improved and impurity-free large-scale synthesis of venlafaxine hydrochloride. Org Proc Res Dev 2011, 15, 1392–5.
[57] Braun JV, Blessing G, Zobel F. Katalytische Hydrierungen unter Druck bei Gegenwart von Nickelsalzen, VI.: Nitrile. Chem Ber 1923, 56, 1988–2001.
[58] Huang Y, Sachtler WMH. On the mechanism of catalytic hydrogenation of nitriles to amines over zeolite supported metal catalysts. Appl Catal, A 1999, 182, 365–78.
[59] Chatterjee A, Shaikh RA, Raj A, Singh AP. Direct reductive hydrolysis of nitriles to aldehydes over Ru- and Pt-loaded zeolites. Catal Let 1995, 31, 301–5.
[60] Huang Y, Sachtler WMH. Intermolecular hydrogen transfer in nitrile hydrogenation over transition metal catalysts. J Catal 2000, 190, 69–74.
[61] Huang Y, Sachtler WMH. Concerted reaction mechanism in deuteration and H/D exchange of nitriles over transition metals. J Catal 1999, 184, 247–61.
[62] Schärringer P, Müller TE, Jentys A, Lercher JA. Identification of reaction intermediates during hydrogenation of CD3CN on Raney-Co. J Catal 2009, 263, 34–41.
[63] Johnson TA, Freyberger DP. Lithium hydroxide modified sponge catalysts for control of primary amine selectivity in nitrile hydrogenations. Chem Ind (Dekker) 2001, 82, 201–27.
[64] Freidlin LK, Sladkova TA. Catalytic reduction of dinitriles. Russ Chem Rev 1964, 33, 319.
[65] Coq B, Tichit D, Ribet S. Co/Ni/Mg/Al layered double hydroxides as precursors of catalysts for the hydrogenation of nitriles: Hydrogenation of acetonitrile. J Catal 2000, 189, 117–28.
[66] Ou EC, Young PA, Norton PR. Interaction of acetonitrile with platinum (111): more properties of the η2(C,N) state and new species in the submonolayer. Surf Sci 1992, 277, 123–31.
[67] Degischer OG, Roessler F, inventors. F. Hoffmann-La Roche A.-G., Switz.; DSM Ip Assets B.V.. assignee. Modification of catalysts for hydrogenation of nitriles to primary amines patent EP1108469A1. 2001.
[68] Ostgard DJ. 2008 Murray Raney Award lecture: The scientific design of activated nickel catalysts for chemical industry. Chem Ind (Boca Raton, FL, U S) 2009, 123, 497–536.
[69] Schärringer P, Müller TE, Lercher JA. Investigations into the mechanism of the liquid-phase hydrogenation of nitriles over Raney-Co catalysts. J Catal 2008, 253, 167–79.
[70] Degischer OG, Roessler F, Rys P. Catalytic hydrogenation of benzonitrile over Raney nickel. Influence of reaction parameters on reaction rates and selectivities. Chem Ind (Dekker) 2001, 82, 241–54.
[71] Krupka J, Drahonsky J, Hlavackova A. Aminocarbene mechanism of the formation of a tertiary amine in nitrile hydrogenation on a palladium catalyst. Reaction Kinetics, Mechanisms and Catalysis 2013, 108, 91–105.
[72] Duch MW, Allgeier AM. Deactivation of nitrile hydrogenation catalysts: New mechanistic insight from a nylon recycle process. Appl Catal, A 2007, 318, 190–8.
[73] Grey RA, Pez GP, Wallo A. Anionic metal hydride catalysts. 2. Application to the hydrogenation of ketones, aldehydes, carboxylic acid esters, and nitriles. J Am Chem Soc 1981, 103, 7536–42.
[74] Fryzuk MD, Montgomery CD, Rettig SJ. Synthesis and reactivity of ruthenium amide-phosphine complexes. Facile conversion of a ruthenium amide to a ruthenium amine via dihydrogen activation and orthometalation. X-ray structure of RuCl(C6H4PPh2)[NH(SiMe2CH2PPh2)2]. Organometallics 1991, 10, 467–73.
[75] Abdur-Rashid K, Faatz M, Lough AJ, Morris RH. Catalytic cycle for the asymmetric hydrogenation of prochiral ketones to chiral alcohols: Direct hydride and proton transfer from chiral catalysts trans-Ru(H)2(diphosphine)(diamine) to ketones and direct addition of dihydrogen to the resulting hydridoamido complexes. J Am Chem Soc 2001, 123, 7473–4.

[76] Takemoto S, Kawamura H, Yamada Y, et al. Ruthenium complexes containing Bis(diarylamido)/ Thioether ligands: Synthesis and their catalysis for the hydrogenation of benzonitrile. Organometallics 2002, 21, 3897–904.

[77] Srimani D, Feller M, Ben-David Y, Milstein D. Catalytic coupling of nitriles with amines to selectively form imines under mild hydrogen pressure. Chem Comm 2012, 48, 11853–5.

[78] Mukherjee A, Srimani D, Ben-David Y, Milstein D. Low-Pressure hydrogenation of nitriles to primary amines catalyzed by ruthenium pincer complexes. Scope and mechanism. Chem Cat Chem 2017, 9, 559–63.

[79] Li H, Hall MB. Computational mechanistic studies on reactions of transition metal complexes with noninnocent pincer ligands: Aromatization–dearomatization or not. ACS Catal 2015, 5, 1895–913.

[80] Bornschein C, Werkmeister S, Wendt B, et al. Mild and selective hydrogenation of aromatic and aliphatic (di)nitriles with a well-defined iron pincer complex. Nat Commun 2014, 5, 4111.

[81] Elangovan S, Topf C, Fischer S, et al. Selective catalytic hydrogenations of nitriles, ketones, and aldehydes by well-defined manganese pincer complexes. J Am Chem Soc 2016, 138, 8809–14.

[82] Mukherjee A, Srimani D, Chakraborty S, Ben-David Y, Milstein D. Selective hydrogenation of nitriles to Primary amines catalyzed by a cobalt pincer complex. J Am Chem Soc 2015, 137, 8888–91.

[83] Tokmic K, Jackson BJ, Salazar A, Woods TJ, Fout AR. Cobalt-catalyzed and lewis acid-assisted nitrile hydrogenation to primary amines: A combined effort. J Am Chem Soc 2017, 139, 13554–61.

[84] Seyden-Penne J. Reductions by the Alumino- and Borohydrides in Organic Synthesis, 2nd edn. New York, Wiley-VCH, 1997.

[85] Schwoegler EJ, Adkins H. Preparation of certain amines. J Am Chem Soc 1939, 61, 3499–502.

[86] Kershaw BJ, Pounder MG, Wilkins KR, inventors. E.I. du Pont de Nemours and Co., assignee. Hexamethylenediamine by catalytic hydrogenation of adiponitrile patent DE2034380A. 1971.

[87] Buehler OR, Keister GP, Long JF, inventors. E.I. du Pont de Nemours and Co.. assignee. Hexamethylenediamine from adiponitrile patent US3461167A. 1969.

[88] Chabert H, inventor. Rhone-Poulenc S. A. assignee. Hydrogenation catalysts for nitriles patent DE2260978A1. 1973.

[89] Nishimura S. Handbook of heterogeneous catalytic hydrogenation for organic synthesis. New York, John Wiley and Sons, 2001.

[90] Raney M, inventor. Catalytic material suitable for use in hydrogenation of oils, etc. US patent US1915473. 1933.

[91] Mullis AM, Bigg TD, Adkins NJ. A microstructural investigation of gas atomized Raney type Al-27.5 at.% Ni catalyst precursor alloys. J Alloys Comp 2015, 648, 139–48.

[92] Freel J, Pieters WJM, Anderson RB. Structure of Raney nickel. I. Pore structure. J Catal 1969, 14, 247–56.

[93] Fouilloux P. The nature of raney nickel, its adsorbed hydrogen and its catalytic activity for hydrogenation reactions (review). Appl Catal 1983, 8, 1–42.

[94] Vedage GA, Armor JN, inventors. Air Products and Chemicals, Inc., USA. assignee. Preparation of secondary amines by disproportionation reaction patent US5574189. 1996.

[95] Cheng WC, Czarnecki LJ, Pereira CJ. Preparation, characterization, and performance of a novel fixed-bed Raney catalyst. Ind Eng Chem Res 1989, 28, 1764–7.

[96] Cheng WC, Lundsager CB, Spotnitz RM, inventors. W. R. Grace and Co., USA. assignee. Shaped Raney catalyst and process for making it patent US4826799A. 1989.

[97] Butler JS. Polyethylene supported catalysts 1985.

[98] O'Hare SA, Mauser JE, Armantrout CE, inventors. The United States of America, assignee. Massive catalyst. US patent US4089812. 1978 May 16, 1978.

[99] Ostgard D, Panster P, Rehren C, Berweiler M, Stephani G, Schneider L, inventors. Degussa AG, assignee. Metal catalysts. US patent US6573213. 2003.
[100] Frank G, Neubauer G, inventors. BASF, assignee. Molded iron catalyst material and its preparation. US patent US4587228. 1986.
[101] Frank G, Rudolf P, Neubauer G, et al., inventors. BASF, assignee. Molded iron catalyst and its preparation. US patent US4521527. 1985.
[102] Jennings JR, inventor. ICI, assignee. Iron catalyst and method of producing it. US patent US4668658. 1987.
[103] Immel O, Langer R, Buysch HJ, inventors. Bayer A.-G, assignee. Process and catalysts for the preparation of (cyclo)aliphatic aminonitriles by hydrogenation of dinitriles. DE patent DE4235466A1. 1994.
[104] Wu JC, inventor. ICI, assignee. Activated iron hydrogenation catalyst. US patent US4480051. 1984.
[105] Adam K, Wimmer K, inventors. BASF, assignee. Verfahren zur Herstellung von partiellen Hydrierungsprodukten des Adipinsaeuredinitrils. DE patent DE848654. 1952.
[106] Dewdney TG, Dowden DA, Morris W, inventors. ICI, assignee. Hexamethylene diamine by hydrogenation of adiponitrile in presence of an activated iron oxide catalyst. US patent US4064172. 1977.
[107] Kershaw BJ, Pounder MG, Wilkins KR, inventors. E. I. du Pont de Nemours and Co., assignee. Hydrogenation of adiponitrile. US Patent US3696153. 1972.
[108] Mul G, Moulijn JA. Preparation of supported metal catalysts (Supported metals in catalysis) In: Anderson JA, Garcia MF, eds. Supported Metals in Catalysis. London, Imperial College Press, 2005, 1–32.
[109] Anderson JA, García MF. Supported Metals in Catalysis. London, Imperial College Press, 2011.
[110] Hutchings GJ, Vedrine JC. Heterogeneous catalyst preparation. In: Baerns M, ed. Basic Principles in Applied Catalysis Berlin, Springer-Verlag, 2004, 215–58.
[111] Regalbuto J. Catalyst Preparation: Science and Engineering. New York, CRC Press, 2016.
[112] Foger K. Dispersed metal catalysts. In: Anderson JR, Boudart M, eds. Catalysis: Science and Technology. Berlin, Heidelberg, Springer Berlin Heidelberg, 1984, 227–305.
[113] Lekhal A, Glasser BJ, Khinast JG. Impact of drying on the catalyst profile in supported impregnation catalysts. Chemical Engineering Science 2001, 56, 4473–87.
[114] Stiles AB, Koch TA. Catalyst manufacture, Second Edition. New York, Marcel Dekker, Inc., 1995.
[115] Munnik P, de Jongh PE, de Jong KP. Recent developments in the synthesis of supported catalysts. Chem Rev 2015, 115, 6687–718.
[116] Boennemann H, Nagabhushana KS. Chemical synthesis of nanoparticles. 2004: American Scientific Publishers, 777–813.
[117] Khajavi H, Stil HA, Kuipers HPCE, Gascon J, Kapteijn F. Shape and transition state selective hydrogenations Using Egg-Shell Pt-MIL-101(Cr) Catalyst. ACS Catal 2013, 3, 2617–26.
[118] Dewhirst KC, inventor. Shell Oil Co. assignee. Homogeneous hydrogenation process employing a complex of ruthenium or osmium as catalyst patent US3454644A. 1969.
[119] Grey RA, Pez GP, Wallo A. Anionic metal hydride catalysts. 2. Application to the hydrogenation of ketones, aldehydes, carboxylic acid esters, and nitriles. J Am Chem Soc 1981, 103, 7536–42.
[120] Beatty RP, inventor. E. I. du Pont de Nemours & Co., USA. assignee. Improved process for reductive hydrolysis of nitriles patent WO9623753A1. 1996.
[121] Beatty RP, Paciello RA, inventors. E. I. du Pont de Nemours & Co., USA. assignee. Process for the preparation of ruthenium complexes and their in situ use as hydrogenation catalysts patent WO9623804A1. 1996.
[122] Beatty RP, Paciello RA, inventors. E. I. du Pont de Nemours & Co., USA. assignee. Preparation of ruthenium cyclohexylphosphine complexes as hydrogenation catalysts patent WO9623803A1. 1996.

[123] Beatty RP, Paciello RA, inventors. E. I. du Pont de Nemours & Co., USA. assignee. Process for the preparation of ruthenium hydrogenation catalysts and products thereof patent WO9623802A1. 1996.
[124] Beatty RP. Homogeneous catalytic hydrogenation and reductive hydrolysis of nitriles. Chem Ind (Dekker) 1998, 75, 183–94.
[125] Beatty RP, inventor. E. I. du Pont de Nemours & Co., USA. assignee. Process for reductive hydrolysis of nitriles patent US5741955A. 1998.
[126] Reguillo R, Grellier M, Vautravers N, Vendier L, Sabo-Etienne S. Ruthenium-catalyzed hydrogenation of nitriles: Insights into the mechanism. J Am Chem Soc 2010, 132, 7854–5.
[127] Enthaler S, Addis D, Junge K, Erre G, Beller M. A general and environmentally benign catalytic reduction of nitriles to primary amines. Chem Eur J 2008, 14, 9491–4.
[128] Gunanathan C, Hölscher M, Leitner W. Reduction of nitriles to amines with H2 catalyzed by nonclassical ruthenium hydrides – Water-Promoted Selectivity for Primary Amines and Mechanistic Investigations. Eur J Inorg Chem 2011, 2011, 3381–6.
[129] Q3D Elemental Impurities: Guidance for Industry. United States Food and Drug Administration. (Accessed October 9, 2017, 2017, at https://www.fda.gov/downloads/drugs/guidances/ucm371025.pdf.)
[130] Chakraborty S, Berke H. Homogeneous hydrogenation of nitriles catalyzed by molybdenum and tungsten amides. ACS Catal 2014, 4, 2191–4.
[131] Chakraborty S, Leitus G, Milstein D. Selective hydrogenation of nitriles to primary amines catalyzed by a novel iron complex. Chem Comm 2016, 52, 1812–5.
[132] Chakraborty S, Milstein D. Selective hydrogenation of nitriles to secondary imines catalyzed by an iron pincer complex. ACS Catal 2017, 7, 3968–72.
[133] Fan L-S. Chapter 1 – Classification and Significance. Gas–Liquid–Solid Fluidization Engineering. Boston, Butterworth-Heinemann, 1989, 3–30.
[134] Bartalini B, Giuggioli M, inventors. Montedison Fibre, assignee. Process for the manufacture of hexamethylenediamine. US patent US3821305. 1974.
[135] Allgeier AM, Koch TA, Sengupta SK. Hydrogenation catalysis in a nylon recycle process. Chem Ind (Boca Raton, FL, U S) 2005, 104, 37–43.
[136] Sengupta SK, Koch TA, Krause KR, inventors. E. I. du Pont de Nemours and Co., assignee. Process for continuous hydrogenation of adiponitrile. US patent US5900511. 1999.
[137] Nedeltchev S, Schumpe A. New Approaches for Theoretical Estimation of Mass Transfer Parameters in Both Gas-Liquid and Slurry Bubble Columns, Mass Transfer in Multiphase Systems and its Applications. In: El-Amin M, ed. Mass Transfer in Multiphase Systems and its Applications, 2011.
[138] Dudukovic MP, Toseland BA, Bhatt BL. A two-compartment convective-diffusion model for slurry bubble column reactors. Ind Eng Chem Res 1997, 36, 4670–80.
[139] Gavroy D, Joly-Vuillemin C, Cordier G, Fouilloux P, Delmas H. Continuous hydrogenation of adiponitrile on Raney nickel in a slurry bubble column. Catal Today 1995, 24, 103–9.
[140] Wilkinson PM, v. Dierendonck LL. Pressure and gas density effects on bubble break-up and gas hold-up in bubble columns. Chem Eng Sci 1990, 45, 2309–15.
[141] Joly-Vuillemin C, Gavroy D, Cordier G, De Bellefon C, Delmas H. Three-phase hydrogenation of adiponitrile catalyzed by raney nickel: Kinetic model discrimination and parameter optimization. Chem Eng Sci 1994, 49, 4839–49.
[142] Mathieu C, Dietrich E, Delmas H, Jenck J. Hydrogenation of adiponitrile catalyzed by Raney nickel. Use of intrinsic kinetics to measure gas-liquid mass transfer in a gas induced stirred slurry reactor. Chem Eng Sci 1992, 47, 2289–94.
[143] Cartolano AR, Vedage GA. Amines by Reduction. Kirk-Othmer Encyclopedia of Chemical Technology. New York, John Wiley & Sons, Inc., 2000.

[144] Machado RM. Increasing productivity in slurry hydrogenation processes. Allentown, PA: Air Products and Chemicals, Inc., 2013.
[145] Xie X, Liotta CL, Eckert CA. CO2-Protected amine formation from nitrile and imine hydrogenation in gas-expanded liquids. Ind Eng Chem Res 2004, 43, 7907–11.
[146] Xie YP, Men J, Li YZ, Chen H, Cheng PM, Li XJ. Catalytic hydrogenation of aromatic and aliphatic nitriles in organic/aqueous biphasic system. Catal Comm 2004, 5, 237–8.
[147] Smith MB, March J. March's Advanced Organic Chemistry: Reactions, Mechanisms, and Structure, 5th edn. John Wiley & Sons, Ltd., 2000.
[148] Bini L, Müller C, Vogt D. Mechanistic studies on hydrocyanation reactions. Chem Cat Chem 2010, 2, 590–608.
[149] Miao X, Dixneuf PH, Fischmeister C, Bruneau C. A green route to nitrogen-containing groups: The acrylonitrile cross-metathesis and applications to plant oil derivatives. Green Chem 2011, 13, 2258–71.
[150] Bezdek MJ, Chirik PJ. Expanding Boundaries: N2 Cleavage and Functionalization beyond Early Transition Metals. Angew Chem – Intl Ed 2016, 55, 7892–6.

Gordon J. Kelly

6 Fischer–Tropsch synthesis – carbon monoxide hydrogenation

6.1 Historical perspective

As we approach the 100-year anniversary of the discovery of the Fischer–Tropsch (FT) synthesis process, it would be fair to say that there have been several peaks and troughs in commercial interest in FT over the decades due to various economic, political, and geographical factors. In 1923, Franz Fischer and Hans Tropsch reported obtaining synthol, a mixture of hydrocarbons and oxygenated compounds, from the reaction of H_2 and CO over alkalized iron at 400–450°C and 150 atmospheres pressure [1]. Soon afterward, Fischer and Tropsch discovered a cobalt-based process that could operate at much lower temperatures and pressures [2]. Initially, cobalt nitrate-impregnated supports were used as catalysts, but soon a coprecipitated Co/ThO_2/kieselguhr catalyst was formulated that would become the standard for many years to come [3, 4]. The first FT plant was brought into in operation on 1936 by Ruhrchemie AG in Oberhausen, Germany, converting coal into liquid fuels (CTL) with a production capacity of 70,000 tons per annum. By 1938, nine FT plants with a combined production capacity of about 500,000 tons per annum were in operation in Germany [5]. Post-WWII investigations into FT processes continued in the United States and Germany [6]:

- Hydrocarbon Research, fluid-bed process, Brownsville, Texas
- Kellog, circulating catalyst process
- Kölbel/Rheinpreußen, slurry process
- Arge (Ruhrchemie-Lurgi) fixed-bed multitubular reactor.

In the mid-1950s, however, the interest in FT processes dwindled with the discovery of abundant oil deposits in the Middle East. The exception to this was South Africa where political and geographic isolation led to the development of a large CTL fuels industry, taking advantage of extremely cheap domestic coal [7]. The first Sasol plant (Sasol I, 500 $bbl.day^{-1}$) was commissioned in 1955 (based on Fe catalyst, Arge fixed bed circulating fluid-bed reactors). This was followed by Sasol II and Sasol III in 1980 and 1982 (initially based on synthol circulating fluid-bed technology, later improved to the Sasol advanced synthol technology, combined capacity in excess of 120,000 $bbl.day^{-1}$ [8]). Interest in FT and other alternative fuels was renewed in the 1970s in the United States and elsewhere due to increasing oil prices and the fear of oil shortages. Major oil companies (Gulf, Exxon, Mobil, Shell, Texaco, and others) and universities began intensive research and development programs in the FT process. During the period (1973–1989),

https://doi.org/10.1515/9783110545210-006

described by Bartholomew as the "Rediscovery of Cobalt" [9], many key catalyst design improvements and characterization techniques were developed. Interest in FT processes again waned in the mid-1980s, however, due to the collapse in world oil prices. In the 1990s, there was another revival of interest in FT as part of gas-to-liquids (GTL) flow sheets to take advantage of "stranded" natural gas reserves and to produce cleaner fuels. In 1993, Shell commissioned a 15,000 $bbl.day^{-1}$ plant in Bintulu, Malaysia (fixed-bed, Co catalyst). Building on the experience of Bintulu, Shell opened the world's largest GTL plant in 2011, Shell Pearl in Qatar which produces 140,000 $bbl.day^{-1}$. In 1992, Mossgas (now PetroSA) commissioned the first large plant using syngas from natural gas and Sasol slurry phase technology (Co catalyst, 20,000 $bbl.day^{-1}$). In 2007, Oryx GTL in Qatar was opened using Sasol's technology with a nameplate production capacity of 34,000 $bbl.day^{-1}$. Currently, another GTL plant using Sasol's technology is undergoing commissioning in Nigeria (Escravos GTL, 34,000 $bbl.day^{-1}$). The drop in the oil price from highs in 2008 of over 130$\$.bbl^{-1}$ to lows of under 30$\$.bbl^{-1}$ in 2016 has lessened the appetite for large GTL plants somewhat, with a number of oil majors cutting back on or closing their GTL projects. With the partial recovery of oil prices, interest continues in GTL (sometimes linked to shale gas opportunities) and CTL processes. In addition to this, FT synthesis is a key step in biomass-to-liquids (BTL) processes to make renewable fuels and municipal solid waste (MSW)-to-liquid fuel concepts. BTL and MSW projects would typically be at a smaller scale (3–5,000 $bbl.day^{-1}$) than GTL and CTL projects.

6.2 Fischer–Tropsch synthesis

The FT synthesis step takes a synthesis gas feed, obtained either by the steam reforming of natural gas or the gasification of coal, biomass, or waste, and converts it to hydrocarbons (primarily linear alkanes and α-olefins). The FT reaction can be considered to be a reductive polymerization of CO by hydrogen, which can be written as follows:

$$n\mathrm{CO} + (2n+1)\mathrm{H_2} \rightarrow \mathrm{C}_n\mathrm{H}_{2n+2} + n\mathrm{H_2O} \quad \text{(for alkanes)}$$

$$n\mathrm{CO} + (2n)\mathrm{H_2} \rightarrow \mathrm{C}_n\mathrm{H}_{2n} + n\mathrm{H_2O} \quad \text{(for alkenes)}$$

In addition to hydrocarbons, water is produced as a coproduct and small amounts of other oxygenated products (alcohols and others). The water formed can react with CO to form CO_2, as a side reaction

$$\mathrm{H_2O} + \mathrm{CO} \rightarrow \mathrm{CO_2} + \mathrm{H_2} \quad \text{(water gas shift)}$$

The amount of water gas shift (WGS) that takes place is linked to the catalyst used; Fe-based catalysts are active WGS catalysts whereas Co catalysts have very low activity for WGS.

FT synthesis is a highly exothermic reaction:

$$CO + 2H_2 \rightarrow -CH_2- + H_2O \quad \Delta H_{298K} = -165\,kJ.mol^{-1}$$

Heat removal and temperature control are therefore key considerations in the choice and design of FT reactors. The FT reaction is operated at high pressure (commonly $P = 20$–45 bar) and there are two operating modes: a high-temperature FT in the range of 300–350°C and low-temperature FT in the range of 200–240°C. This temperature difference, together with the catalyst choice, determines the product distribution [10].

As the reaction is a polymerization process, it should be borne in mind that FT synthesis produces a spectrum of products such that any particular carbon chain length product can only achieve a certain maximum in selectivity. The product distribution of hydrocarbons formed during the FT process follows an Anderson–Schulz–Flory (ASF) distribution [11] which can be expressed as

$$\frac{W_n}{n} = (1-\alpha)^2 \alpha^{n-1}$$

where W_n is the weight fraction of hydrocarbons containing n carbon atoms and α is the chain growth probability or the probability that a molecule will continue reacting to form a longer chain rather than terminating. Taking the natural log form of the equation gives

$$\ln\left(\frac{W_n}{n}\right) = n \ln \alpha + \ln\left(\frac{(1-\alpha)^2}{\alpha}\right)$$

A plot of $\ln(W_n/n)$ versus n should produce, for an ASF distribution, a straight line with a gradient of $\ln \alpha$. In practice, there can be marked deviations from the ASF model, common deviations found include

- CH_4 made in a greater quantity than predicted
- A lower amount of C_2s (ethane/ethene) formed
- A change in the chain growth probability (α) with carbon number

Figure 6.1 shows a typical alpha plot for a supported Co catalyst. Cobalt catalysts typically demonstrate higher than predicted methane makes and lower C_2 makes during FT synthesis.

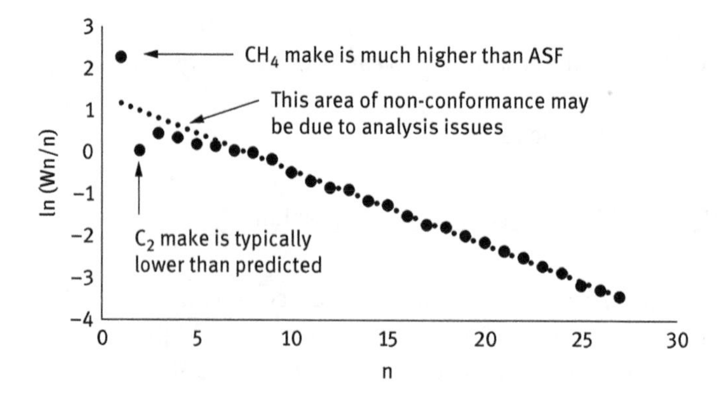

Figure 6.1: Typical alpha plot for a supported Co catalyst.

Selectivity limitations are inherent in the chain growth mechanism for FT synthesis which is governed by ASF kinetics [11].

Although the product molecular weight can be varied by choosing process conditions and/or catalysts to achieve a given degree of polymerization (α in Figure 6.2), a wide distribution of products is inherent in the process. For instance, the maximum obtainable weight percentage of light LPG (liquid petroleum gas) hydrocarbons (C_2–C_4) is 56%, of gasoline (C_5–C_{11}) is 47%, and of diesel fuel (C_{12}–C_{17}) is 25%.

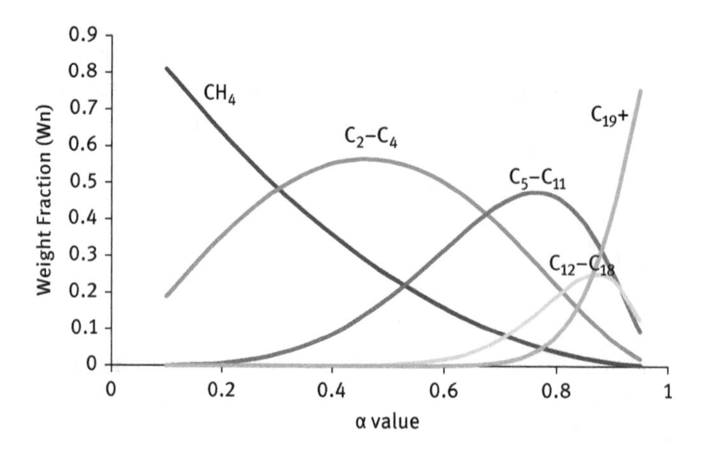

Figure 6.2: Weight fraction of hydrocarbon products as a function of chain growth probability (α).

At short carbon chain lengths, a high level of α-olefins can be produced during FT synthesis. The α-olefin selectivity, often expressed by a paraffin/olefin (P/O) ratio, generally increases exponentially with chain length $n > 2$. For $n = 2$, the P/O ratio is typically significantly deviated from the curve (Figure 6.3).

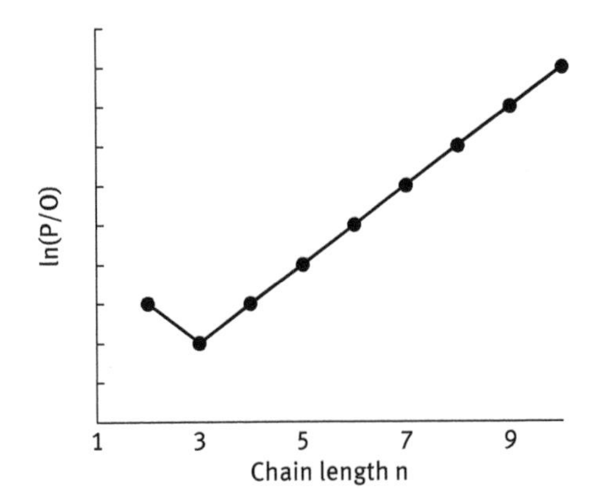

Figure 6.3: Schematic illustration of the chain length dependence of *P*/*O* ratio in FT synthesis [12].

Fe catalysts can be modified with promoters that limit hydrogenation reactions. These catalysts can be operated to produce high levels of C_2–C_4 olefins with *P*/*O* ratios approaching 0.1 [13]. Under these conditions, low levels of methane can be produced, lower than what would be predicted from ASF kinetics. As the products in this case are principally olefins, the C_1 production is suppressed as you cannot form a C_1 olefin.

The wide range of hydrocarbon products that are produced during FT synthesis (gases, liquids, and waxes) can make the exit analysis extremely complex. The aqueous phase produced can also contain small amounts of alcohols, aldehydes, acids, and ketones. Figure 6.4 shows a typical GC analysis of the wax produced from the slurry phase test of a Co catalyst.

In many FT publications, to simplify the results in addition to the catalyst rate, the following selectivities are typically reported:

- Lights — $\%CH_4$, $\%C_2$–C_4, $\%CO_2$
- (*P*/*O*) ratio — at a specified chain length
- Products — $\%C_{5+}$

The C_{5+} figure is calculated by carrying out accurate online analysis of the lights and then assuming that any other conversion products are $\geq C_5$. For consistency, catalysts selectivities should be quoted at the same conversion level. C_{5+} figures can be checked against α figures calculated from off-line wax analysis.

A low selectivity to methane during FT is a key factor in the process economics of GTL processes. To suppress methane production to an economically viable level, high α figures are required ($\alpha > 0.9$); this will produce a high percentage of long-chain hydrocarbons and waxes. The wide range of hydrocarbons that are produced in the FT process undergoes separation and further treatments such as hydrocracking and hydroisomerization to produce the required product slate. A simplified process flow sheet for GTL and CTL is shown in Figure 6.5.

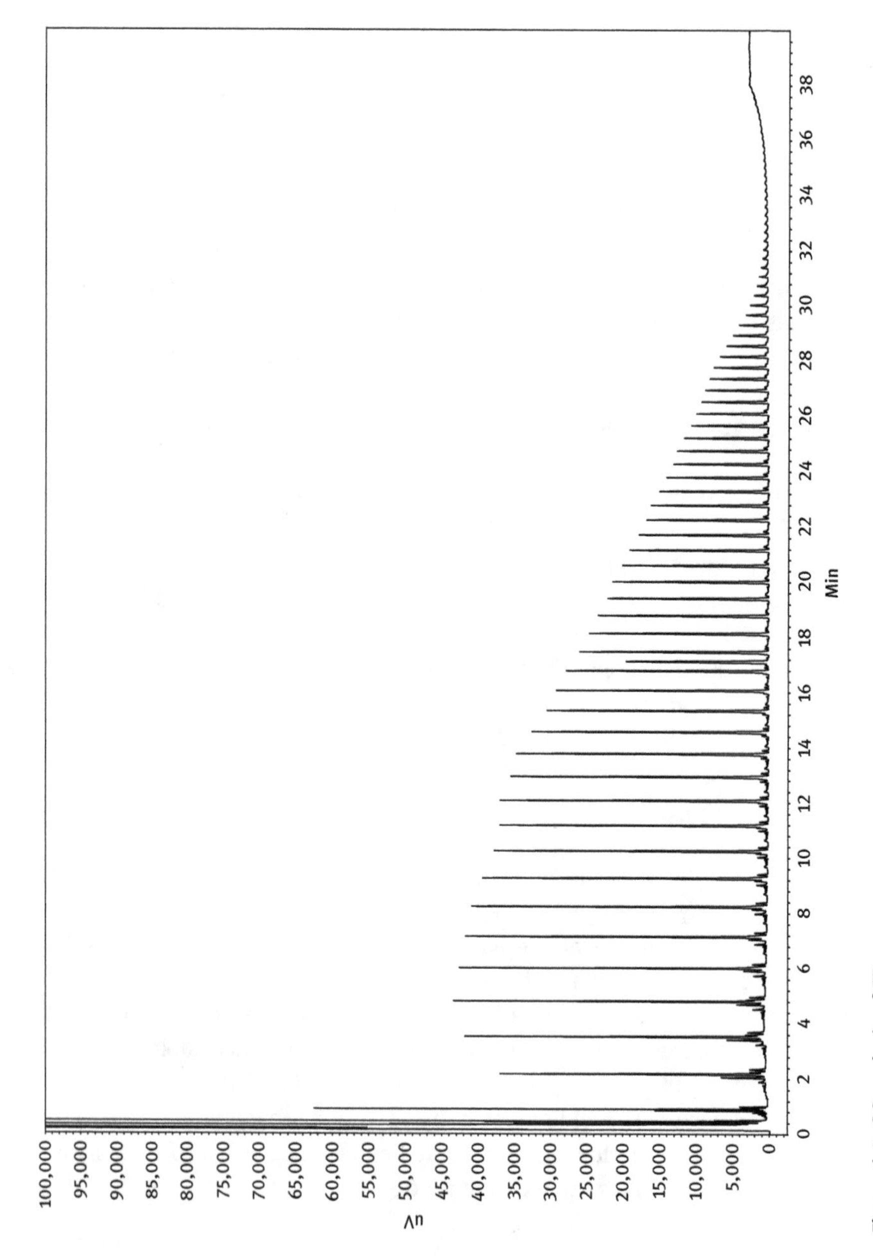

Figure 6.4: GC analysis of FT wax.

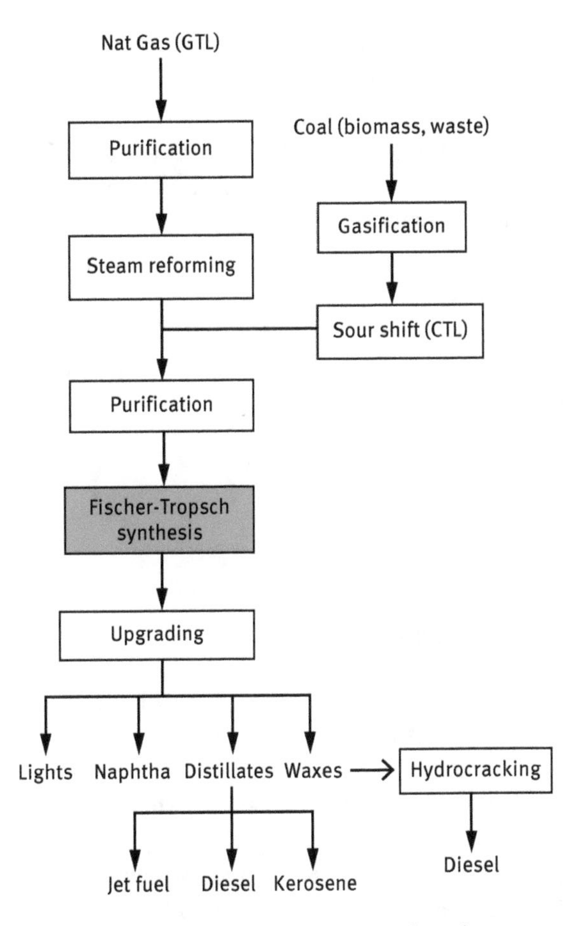

Figure 6.5: Simplified GTL and CTL process flow sheet.

The ability to exploit FT technology profitably is dependent on the oil price and the cost of the natural gas or feedstock used. FT technology has the potential to unlock stranded gas resources and unlock oil resources that would otherwise not be produced unless associated gas is converted. The products formed by FT synthesis also have some particular advantages over products derived from crude oil. They are, for instance, totally free of sulfur, nickel, vanadium asphaltenes, and aromatics that are typically found in crude oil. The FT products are almost exclusively paraffins and α-olefins with very few or no complex cyclic hydrocarbons or oxygenates. The main attraction of FT diesel relates to its purity (no detectable sulfur or aromatics) and high cetane values. Testing has shown that FT diesel fuels can provide reductions in engine emissions of particulate matter, CO, and NO_x [14].

6.3 Fischer–Tropsch catalysts overview

The four metals most studied for catalyzing the FT reaction are Ni, Fe, Co, and Ru. Of these four, Ni is not considered as an industrial option due to its high selectivity to methane. Ru tends to be discounted as an industrial option because of scarcity and cost. Figure 6.6 summarizes the typical selectivity behavior displayed by the Group VII metals during CO hydrogenation.

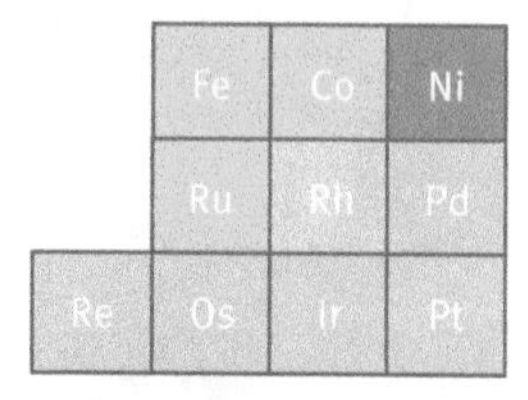

Suitable for FT synthesis
Produces methane
Yield mostly oxygenated products
Produces C_2 and higher oxygenates and hydrocarbons

Figure 6.6: CO hydrogenation selectivity over Group VIII metals.

The choice between using Fe- and Co-based catalysts is not so clear cut but in the simplest terms, the ratio of H_2:CO in the syngas feed can determine the choice of catalyst. In broad terms, Fe catalyst technology can be preferred for coal and refinery bottom-based processes (lower H_2:CO ratios), whereas Co-based systems are the preferred technology for converting natural gas to FT naphtha (higher H_2:CO ratios). Fe catalysts are used in both high-temperature (300–350°C) and low-temperature (200–240°C) processes [15]. The high-temperature process produces C_1–C_{15} hydrocarbons and a large range of valuable chemicals such as α-olefins and oxygenates. The low-temperature process targets the production of linear long-chain hydrocarbons. The high-temperature Fe catalyst is a fusion of iron oxide together with the chemical promoter K_2O and structural promoters such as MgO or Al_2O_3. The low-temperature Fe catalyst is prepared by a precipitation method and has a Fe/Cu/K/SiO_2 formulation [15]. Co FT catalysts usually contain between 15 wt% and 30 wt% cobalt. Often these catalysts contain small amounts of a second metal promoter (typically noble metals) and oxide promoters (zirconia, lanthia, and cerium oxide). The active phase is usually supported by an oxide with a high surface area (silica, alumina, and titania) [16]. The catalytic properties of cobalt-based catalyst are highly influenced by the support material used as the reducibility and dispersion of the cobalt are influenced by the textural and surface properties of support.

Table 6.1 compares the typical characteristic features of low-temperature Fe and Co FT catalysts, while Figure 6.7 demonstrates the typical differences in selectivity for Co, Fe, and Ru catalysts during FT synthesis. The catalysts were all tested at 210°C, 20 bar, and with a H_2:CO ratio of 2. Under these conditions, the high CO_2 make of the Fe

Table 6.1: Low-temperature Fe and Co FT catalyst characteristics.

	Fe	Co
Typical catalyst properties	Low cost Lower activity Silica support Shorter lifetimes Promoted with Cu, K_2O Active phase – Fe carbides	Higher cost Higher activity Supported (Al_2O_3, SiO_2, TiO_2) Longer lifetime (~2 years) Oxide and noble metal promoters Active phase – Co metal
Catalyst Deactivation	Can tolerate S levels of ~2 ppm Strong tendency to form carbon and deactivate. Carbon laydown can also lead to catalyst particle expansion and breakage	Requires S levels to be less than 5 ppb Less tendency to form carbon More easily regenerated *in-situ*
Hydrogenation activity	Lower hydrogenation activity Produces less methane Produces more olefinic products Produces more oxygenated species (alcohols and aldehydes)	Higher hydrogenation activity Produces more methane Produces more paraffinic products Produces less oxygenated species (alcohols and aldehydes)
Water gas shift activity	Good WGS catalyst Can operate at low H_2:CO ratios Has lower C efficiency Some oxygen ejected as CO_2	Poor WGS catalyst Operates at higher H_2:CO ratios Has higher C efficiency Water is the principal oxygenate product

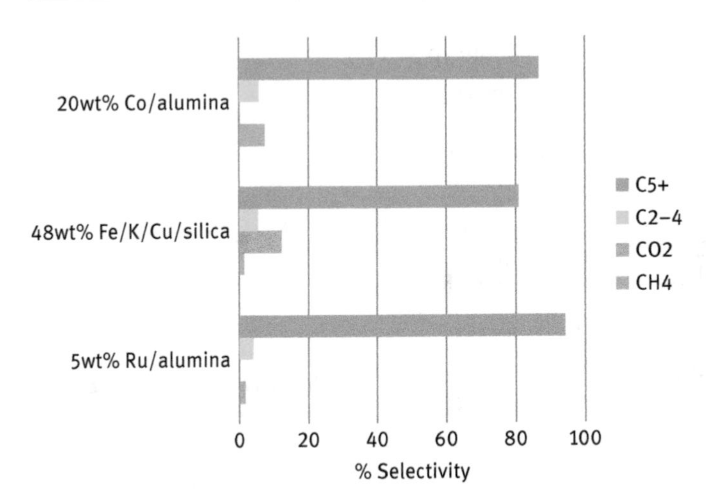

Figure 6.7: Typical selectivities of Co, Fe, and Ru Fischer–Tropsch catalysts.

catalysts and the higher methane make of the Co catalysts are clearly apparent. The relative activities of these three catalysts are as follows: 20 wt% Co/alumina = 1, 5 wt%

Ru/alumina = 0.51, 48 wt% Fe/K/Cu/silica = 0.40. Ru catalysts can produce very high C_{5+} selectivities but their cost still makes their use commercially prohibitive.

6.4 FT mechanism and kinetics

The mechanism of FT synthesis is still a matter of major debate. Three general types of mechanism have been proposed in the literature; the carbene mechanism, the hydroxyl-carbene mechanism, and CO insertion mechanism. The carbene mechanism, proposed by Fischer and Tropsch, suggested that the carbon chain propagated via the stepwise polymerization of CH_2 intermediates on the metal surface [1–3]. The hydroxyl-carbene mechanism proposed by Anderson and Emmett progresses via the dimerization of hydroxyl methylene intermediates [17]. Pichler and Schulz suggested that the chain growth progressed through the insertion of CO into adsorbed alkyl intermediates [18]. Of the three mechanisms, the carbene mechanism (Figure 6.6) is the most commonly supported by experimental evidence and theoretical work. In this mechanism, the C–O bond breaks and forms surface-adsorbed C and O species. These are both sequentially hydrogenated to give C_1 surface species (CH_x, $x = 1–3$) and water. The C_1 species can either be hydrogenated to form CH_4 or take part in C–C coupling reactions to form C_2 and longer chain hydrocarbons.

Various kinetic equations for FT synthesis over Co catalysts have been derived [15]. A suitable rate equation that is often cited is as follows:

$$\text{Rate} = \frac{aP_{CO}P_{H_2}}{(1 + bP_{CO})^2}$$

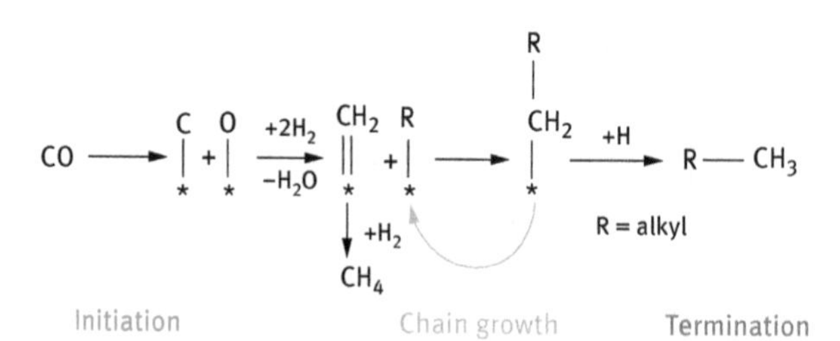

Figure 6.8: Fischer–Tropsch carbene mechanism.

Other forms of the rate equation have been derived which include an inhibition effect by water and a P_{H_2O} factor in the denominator of the equation. Equations of this form may be appropriate where the catalyst contains small Co crystals (<5 nm) which are thought to be vulnerable to oxidation at high partial pressures of water. Activation energies of 93–95 $kJ.mol^{-1}$ have been reported for the reaction [19].

6.5 FT reactors and processes

The choice of reactor depends upon several important attributes that include
1. size/throughput
2. capital and operating costs
3. thermal efficiency
4. heat removal
5. product selectivity
6. flexibility in terms of operating conditions and product quality
7. maintenance of catalyst activity and/or ease of regeneration
8. reactor ideality and/or stability

A range of reactor types has been used or proposed for use for FT synthesis. The four main reactor types which have been used commercially [fixed-bed, fluidized-bed (circulating), fluidized-bed (fixed), and slurry-bed reactors] are compared in Table 6.2. More recently, advanced reactors and micro-channel reactor technologies have been investigated to improve the heat transfer and to intensify the process. These technologies are often targeted at smaller, stranded gas fields or BTL or MSW projects.

6.6 Cobalt Fischer–Tropsch catalysts

Although cobalt is more expensive than iron, it is the metal of choice for the synthesis of long-chain hydrocarbons due to its higher productivity and longer life. Studies by Iglesia demonstrated that the FT reaction was structure-insensitive over supported Co catalysts [20], with the turnover frequency (TOF) for the reaction being independent of the Co particle size when the Co particles were in the size range 10–100 nm (Figure 6.9).

Recent work by de Jong et al. [21] has shown that Co particles of <6 nm demonstrate lower TOF than particles >6 nm (Figure 6.10). From steady-state isotopic transient kinetic analysis, it was concluded that the drop in TOF in particles <6 nm was the result of a significant increase in the CH_x residence time combined with a decrease in CH_x coverage [21].

Various deactivation mechanisms have been reported for Co FT catalysts, including Co particle sintering, poisoning, carbon fouling, and oxidation of cobalt to cobalt oxide or cobalt aluminate (for alumina-supported catalysts). The loss of metal cobalt due to attrition is also an issue for three-phase slurry reactors. Deactivation of Co

Table 6.2: Fischer–Tropsch reactor types advantages and disadvantages.

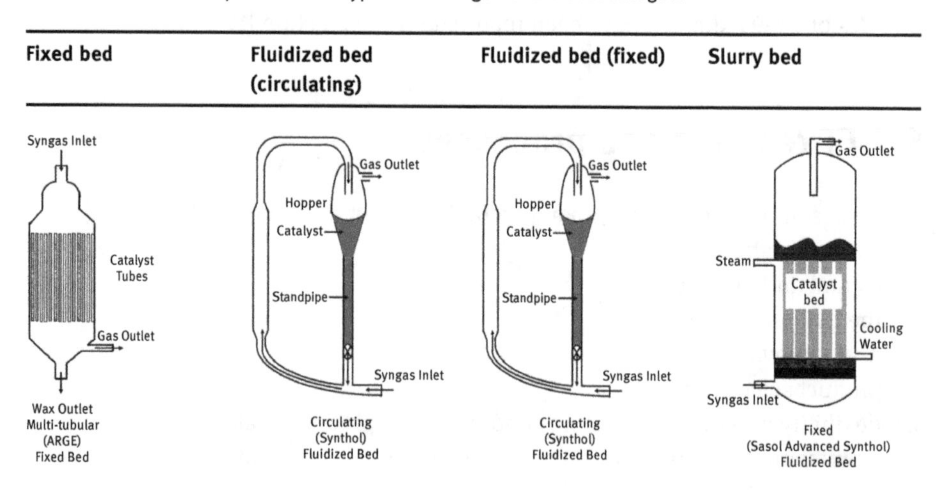

Fixed bed	Fluidized bed (circulating)	Fluidized bed (fixed)	Slurry bed
Multitubular with water as heat transfer medium.	Finely divided catalyst and gas flow upward co-currently in a pipe reactor at high velocity.	Stationary bed with internal heat exchanger.	Three-phase reactor. Solid catalyst phase suspended in liquid phase through which the reactant gases are passed.
Advantages Simple to operate. Can be used over a wide temperature range irrespective of whether the FT products are gaseous or liquids. No catalyst separation issue“Slugs” of H_2S are adsorbed on top layer of catalyst.	**Advantages** Efficient heat transfer characteristics. Higher gas throughputs. Lower pressure dropFresh catalysts can be added without process disturbance.	**Advantages** Cheaper to construct. Less erosion/Less maintenance. Better heat removal. Steadier higher conversion. Near isothermal operation. Lower pressure. Lower compression costs.	**Advantages** Simple construction. Lower capital/operating costs. Very efficient heat transfer and uniform temperature. High catalyst efficiency. Low pressure drop. Higher catalyst loadings possible. High single pass conversion. Favorable product distribution. Low methane. No-bed plugging. Can operate at higher reaction temperatures. Capable of running at low H_2:CO ratios without C deposition.

Table 6.2: (*continued*)

Fixed bed	Fluidized bed (circulating)	Fluidized bed (fixed)	Slurry bed
Disadvantages Reaction transport limited => Small particles are required which leads to high pressure drops and high compression costs. Long downtime required to load and unload tubes. Heat transfer problems can lead to hot spots.	**Disadvantages** More complex to operate. Erosion can occur at areas of high velocity. Separation of catalyst fines is required from exhaust gas. Carbon deposition can lead to a conversion decline.	**Disadvantages** Higher reaction temperature required. Higher selectivity for lighter products. Catalyst requires greater resistance to attrition.	**Disadvantages** Attrition of catalysts. Catalyst separation issues. Difficult to model. Scale-up issues.

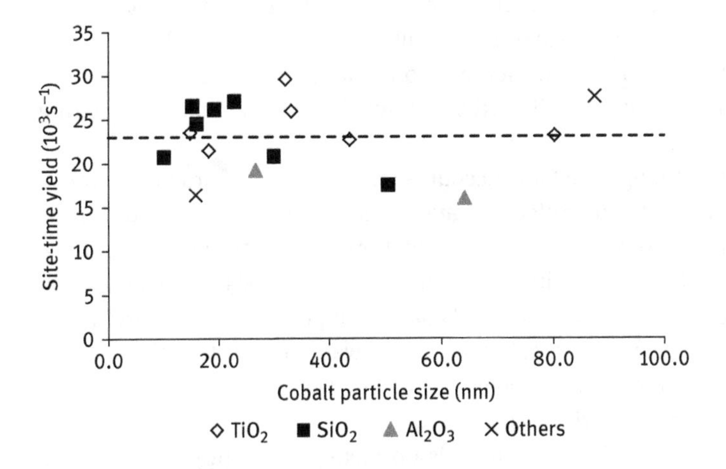

Figure 6.9: Effect of Co particle size and support on FT turnover frequency (200°C, 20 bar, $H_2/CO = 2.1$).

catalysts through oxidation appears to be a function of the reactor partial pressure of hydrogen and water and cobalt particle size [22]. Thermodynamic calculations suggest that the oxidation of Co catalysts is only possible on very small cobalt particles (<5 nm) [22]. Cobalt particles of <5 nm, when supported on metal oxide carriers such as alumina, tend to have high metal support interactions which can make reduction difficult [23]. Smaller cobalt crystal sizes have also been linked to higher methane selectivities during FT [22, 24]. Studies where additional H_2O was introduced to the FT reaction concluded that a cobalt cluster size of greater than 10 nm was beneficial to stabilizing the catalyst [25]. The optimum cobalt particle size for supported cobalt FT

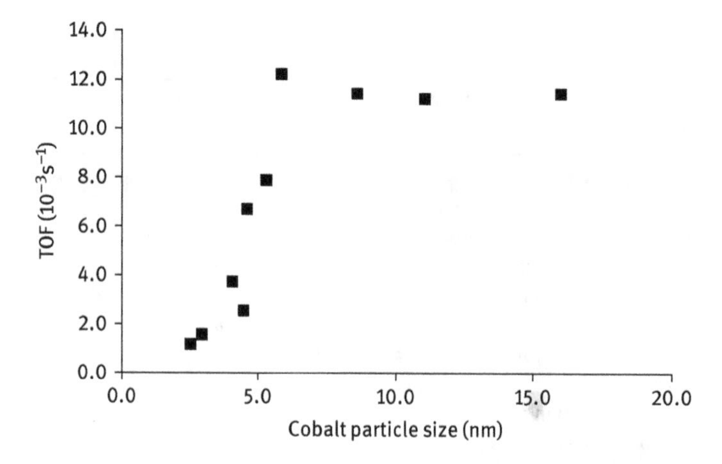

Figure 6.10: FT activity (1 bar, 220°C, $H_2/CO = 2$) as a function of cobalt particle size. Reprinted (adapted) with permission from Ref. [21]. Copyright (2009) American Chemical Society.

catalysts is therefore a compromise between a number of factors; small well-dispersed cobalt particles will give high FT activity but will be susceptible to oxidation, support interactions, sintering, methanation reactions, and a lower intrinsic TOF at less than 7–8 nm; larger cobalt particles should be more resistant to sintering but they will have a lower initial activity.

Water is the main FT by-product and accounts for ca. 50 wt% of products formed. The interaction of water with supported Co catalysts is complex and depends on the support and its nature, Co metal loading, promotion with noble metals, and the preparation procedure [26]. The addition of water in FT synthesis can lead to both increases and decreases in CO conversion and changes in product selectivities. Cobalt particle size, reactor partial pressure of hydrogen and water, the surface chemistry of the support, and the average pore diameter of the support all appear to be factors which influence the interaction with water [26].

The active phase for cobalt FT catalysts is Co metal, therefore prior to use Co catalysts must be activated by reduction. Commercially supported cobalt catalysts for FT synthesis typically contain a reduction promoter. Reduction promoters that are commonly utilized include Pt, Ru, and Re. Debate continues on the function and location of reduction promoters [27]. Reported effects of reduction promoters include the following:

- Improved Co dispersions achieved during preparations
- Decreased Co reduction temperature
- Increased Co reducibility (degree of reduction)

The reported effects on performance are, however, less consistent [27]:

- Increased TOF/TOF not changed
- Improved stability/Faster deactivation

- Improved C_{5+} and lower methane/No effect on selectivity
- Increased oxygenate production/No effect on oxygenates

Temperature-programed reduction (TPR) is a widely used tool for the characterization of the reduction process of supported Co catalysts. The TPR method yields information on the reducibility of the material and surface interactions. Figure 6.11 shows a typical TPR profile for a non-promoted 20 wt% Co/γ-alumina catalyst. The two main peaks are from the reduction of Co_3O_4 to CoO followed by the reduction of CoO to Co metal. Smaller peaks may be present from residual nitrates from the catalyst preparation step and, at higher temperature, due to interactions between the Co and the support.

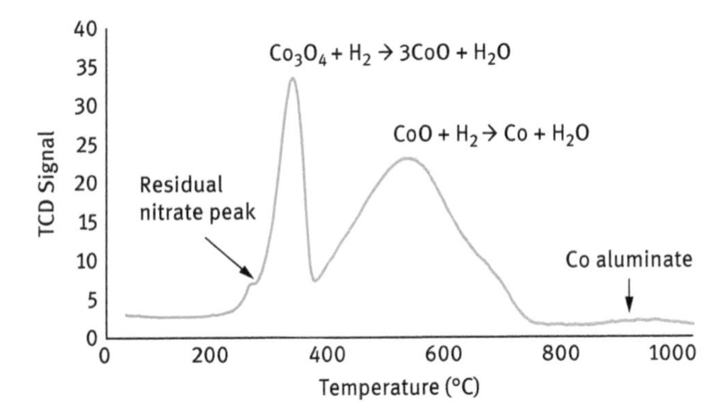

Figure 6.11: TPR of a non-promoted 20 wt% Co/γ-alumina catalyst. [Temperature programed reduction (TPR) experiments were carried out on an Altamira AMI-200 instrument using 10% H2 in argon].

Figure 6.12 shows the effect of the addition of Pt (0.0005–5 wt%) as a reduction promoter to a 20-wt% Co/γ-alumina catalyst prepared by incipient wetness impregnation. The most rapid change in the TPR profile is between 0.01 wt% and 0.1 wt% Pt. Assuming an average cobalt crystal diameter of 10 nm and assuming that the Pt promoter is all associated with the cobalt phase, this gives a range between 3 and 30 Pt atoms per cobalt crystal.

6.7 Iron Fischer–Tropsch catalysts

Iron-based Fischer catalysts can be considered for coal-based processes due to several positive features [28]:
- Low cost when compared to cobalt
- Feed flexibility (H_2/CO = 0.5–2.5)
- High-temperature adaptability (230–350°C)

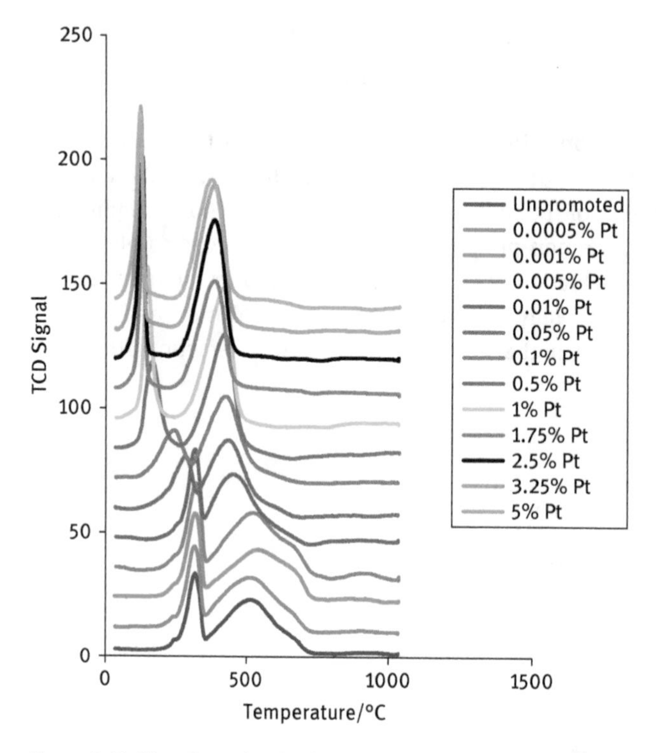

Figure 6.12: The effect of Pt loading on reduction temperature.

- High FT activity
- Some resistance to poisoning (in comparison to Co catalysts)

For Fe FT catalysts, there is less clarity about the nature of the active site, however, than there is for Co catalysts. *In-situ* Mossbauer spectroscopy experiments [29] have linked FT activity with the formation of various iron carbides (FeC_x species). Other researchers suggest that Fe_2O_3 is the active phase [30]. The complexity of the commercial Fe-based FT catalysts in terms of metals, supports, and promoters does not lend themselves to the clear identification of the active site. Research groups have reported preferred activation procedures with carbon monoxide [31], hydrogen [32], or synthesis gas [33].

References

[1] Fischer, F., Tropsch, H. The Preparation of Synthetic Oil Mixtures (synthol) from Carbon Monoxide and Hydrogen, Brennstoff-Chem. 1923, 4, 276.
[2] Fischer, F., Tropsch, H. The Synthesis of Petroleum at Atmospheric Pressures from Gasification Products of Coal, Brennstoff-Chem. 1926, 7, 97.

[3] Fischer, F., Koch, H. New Developments in the Adaptation of Cobalt Catalysts to the Benzine Synthesis, Brennstoff-Chem. 1932, 13, 61.

[4] Pichler, H. Twenty-five Years of Synthesis of Gasoline by Catalytic Conversion of Carbon Monoxide and Hydrogen, Adv. Catal. 1952, 4, 271.

[5] Casci, J.L., Lok, C.M. and Shannon, M.D. Fischer–Tropsch Catalysis: The Basis for an Emerging Industry with Origins in the Early 20th Century, Catal. Today 2009, 1 45, 38–44.

[6] Schulz, H. Short History and Present Trends of Fischer–Tropsch Synthesis, Appl. Catal. A Gen. 1999, 186, 3–12.

[7] https://www.netl.doe.gov/research/coal/energy-systems/gasification/gasifipedia/history-gasification, accessed 11/08/2017

[8] Steynberg, A.P. Chapter 1 – Introduction to Fischer-Tropsch Technology, Stud. Surf. Sci. Catal., 2004, 152, 1–63.

[9] Bartholomew, C.H. Presentation at the National Spring Meeting of the American Institute of Chemical Engineers, New Orleans, March 30 – April 3, 2003.

[10] Dry, M.E. The Fischer-Tropsch Process: 1950–2000, Catal. Today 2002, 71, 227–241.

[11] Anderson, R.B. Catalysts for the Fischer-Tropsch Synthesis, Vol. 4. Van Norstarnd Reinhold, New York, 1956.

[12] Song, C.J., Hu, P., Lok, C.M., Ellis, P., French, S. A Density Functional Theory Study of the α-olefin Selectivity in Fischer–Tropsch Synthesis, J. Catal. 2008, 255, 20.

[13] Torres Galvis, H.M., Bitter, J.H., Khare, C.B., Ruitenbeek, M., Dugulan, I. and de Jong, K.P. Supported Iron Nanoparticles as Catalysts for Sustainable Production of Lower Olefins, Science 2012, 335(6070),835–838.

[14] Gowdagiri, S., Cesari, X.M., Huang, M. and Oehlschlaeger, M.A. Investigation of the Effects of Biodiesel Feedstock on the Performance and Emissions of a Single-Cylinder Diesel Engine, Fuel 2014, 136, 253–260.

[15] Steynberg, A.P. and Dry, M.E. (Eds.), Fischer-Tropsch Technology, Stud. Surf. Sci. Catal. 2007, 152, 1–722.

[16] Khodakov, A.Y. Fischer-Tropsch Synthesis: Relations Between Structure of Cobalt Catalysts and their Catalytic Performance, Catal. Today 2009, 144, 251–257.

[17] Kummer, J.F. and Emmett, P.H. Fischer – Tropsch Synthesis Mechanism Studies. The Addition of Radioactive Alcohols to the Synthesis, Gas J. Am. Chem. Soc. 1953, 75, 5177.

[18] Pichler, H. and Schulz, H. New Insights in the Area of the Synthesis of Hydrocarbons from CO und H2, Chem. Ing. Tech. 1970, 12, 1160.

[19] Yates, I.C. and Satterfield, C.N. Intrinsic Kinetics of the Fischer-Tropsch Synthesis on a Cobalt Catalyst, Energy Fuels 1991, 5(1),168–173.

[20] Iglesia, E., Reyes, S.C., Madon, R.J. and Soled, S.L. Selectivity Control and Catalyst Design in the Fischer-Tropsch Synthesis: Sites, Pellets, and Reactors, Adv. Catal., 1993, 39, 221–301.

[21] den Breejen, J.P., Radstake, P.B., Bezemer, G.L., Bitter, J.H., Frøseth, V., Holmen, A. and de Jong, K.P. On the Origin of the Cobalt Particle Size Effects in Fischer-Tropsch Catalysis, J Am Chem Soc, 2009, 131, 7197–7203.

[22] Van de Loosdrecht, J., Balzhinimaev, B., Dalmon, J.A., Niemantsverdriet, J.W., Tsybulya, S.V., Saib, A.M., van Berge, P.J. and Visagie, J.L. Cobalt Fischer-Tropsch synthesis: Deactivation by oxidation? Catal. Today 2007, 123, 293–302.

[23] Vogel, A.P., van Dyk, B., Saib, A.M. GTL Using Efficient Cobalt Fischer-Tropsch catalysts, Catal. Today 2016, 259–323.

[24] Borg, Ø., Dietzel, P.D.C., Spjelkavik, A.I., Tveten, E.Z., Walmsley, J.C., Diplas, S., Eri, S., Holmen, A. and Rytter, E. Fischer–Tropsch Synthesis: Cobalt Particle Size and Support Effects on Intrinsic Activity and Product Distribution, J. Catal. 2008, 259, 161.

[25] Jacobs, G., Das, T.K., Patterson, P.M., Luo, M., Conner W.A. and Davis, B.H. Fischer–Tropsch Synthesis: Effect of Water on Co/Al2O3 Catalysts and XAFS Characterization of Reoxidation Phenomena, Appl. Catal. A: Gen. 2004, 270, 65.
[26] Dalia A.K. and Davis, B.H. Fischer–Tropsch Synthesis: A Review of Water Effects on the Performances of Unsupported and Supported Co Catalysts, Appl. Catal. A: Gen, 2008, 348(1),1–15.
[27] Ma, W., Jacobs, G., Keogh, R.A., Bukur, D.B. and Davis B.H., Fischer–Tropsch synthesis: Effect of Pd, Pt, Re, and Ru Noble Metal Promoters on the Activity and Selectivity of a 25% Co/Al_2O_3 Catalyst, Appl. Catal. A: Gen 2012, 1–9, 437–438.
[28] Zhang, J., Abbas, M. and Chen, The Evolution of Fe Phases of a Fused Iron Catalyst during Reduction and Fischer–Tropsch Synthesis, J. Catal. Sci. Technol. 2017, 7, 3626.
[29] Motjope, T.R., Dlamini, H.T., Hearne, G.R. and Coville, N.J., Application of in situ Mössbauer Spectroscopy to Investigate the Effect of Precipitating Agents on Precipitated iron Fischer–Tropsch catalysts, Catal. Today 2002, 71, 335–341.
[30] Butt, J.B., Carbide Phases on iron-Based Fischer-Tropsch Synthesis Catalysts Part I: Characterization Studies, Catal. Lett. 1990, 7, 61–82.
[31] O'Brien, R. J., Xu, L., Milburn, D. R., Li, Y.-X., Klabunde, K. J. and Davis, B. H., Fischer-Tropsch Synthesis: Impact of Potassium and Zirconium Promoters on the Activity and Structure of an Ultrafine Iron oxide Catalyst, Top. Catal. 1995, 2, 1.
[32] Bukur, D. B., Koranne, M., Lang, X., Rao, K. R. P. M., Huffman, G. P. Pretreatment Effect Studies with a Precipitated Iron Fischer-Tropsch Catalyst, Appl. Catal. 1995, 126, 85.
[33] Kölbel, H., Ralek, M. The Fischer-Tropsch Synthesis in the Liquid Phase Catal. Rev.-Sci. Eng. 1980, 21, 225.

Justin S. J. Hargreaves

7 Heterogeneously catalyzed ammonia synthesis

7.1 Introduction

The development of the Haber–Bosch process was undoubtedly a landmark of the twentieth century. Through the provision of access to synthetic fertilizers, this single process has been directly credited with sustenance of a significant fraction of the global population. Indeed, nitrogen fixation from natural means accounts for about 50% of that necessary for plant and crop growth with the Haber–Bosch process accounting for the remainder [1, 2]. This is a fact which is reflected in the nitrogen content within the bodies of each and everyone of us in that a significant proportion will have passed through Haber–Bosch process. Further statistics associated with the process are that it is estimated, when considered in its entirety, to be responsible for the consumption of 1–2% of manmade energy [3] and to be responsible for a significant proportion of the global manmade carbon dioxide emissions. In terms of the latter point, the CO_2 intensive nature of the process arises in the use of fossil fuel sources to derive hydrogen and in their application as energy source. In 2010, the process was operated at the scale of 131 million tons of ammonia production, releasing an associated 245 million tons of CO_2 [4].

In general industrial operation, the process involves the reaction between nitrogen and hydrogen over a promoted iron-based catalyst. The feedstreams have to be of high purity as the catalyst is very susceptible to poisoning by even very small traces of oxygenates. As can be seen below, the reaction is favored by high pressure, and reaction pressures of >100 atmospheres are applied on the industrial scale.

$$1/2N_2 + 3/2H_2 \leftrightarrow NH_3 \quad \Delta H^\circ = -46\,\text{kJ.mol}^{-1}$$

Although the reaction is thermodynamically favored at low reaction temperatures, ca. 400–500°C is applied in practice to achieve acceptable process kinetics. Table 7.1, taken from Fritz Haber's Nobel Prize lecture [5], presents the thermodynamic ammonia yields as a function of applied pressure and temperature.

Detailed descriptions of the process and the catalyst applied can be found in a number of monographs, e.g., Ref. [6–8]. In the iron-based catalytic system, a pronounced degree of structure sensitivity is observed with C7 sites believed to be the active site [9]. Nitrogen activation is considered to be the rate-determining step and the potassium dopants applied are believed to facilitate this via electronic and structural promotion [10]. Fundamental insights into the nature of iron-catalyzed ammonia synthesis which provide insight into the reaction steps have been detailed in the

https://doi.org/10.1515/9783110545210-007

Table 7.1: The percentage of ammonia at equilibrium for a 3:1 H_2/N_2 mixture as a function of temperature and pressure.

Temperature (°C)	1 atm	30 atm	100 atm	200 atm
200	15.3	67.6	80.6	85.8
300	2.18	31.8	52.1	62.8
400	0.44	10.7	25.1	36.3
500	0.129	3.62	10.4	17.6
600	0.049	1.43	4.47	8.25
700	0.0223	0.66	2.14	4.11
800	0.0117	0.35	1.15	2.24
900	0.0069	0.21	0.68	1.34
1000	0.0044	0.13	0.44	0.87

Table adapted from Ref. [5].

work of Ertl and coworkers, e.g., Ref. [11]. The development of the industrial catalyst in the context of early research undertaken at BASF has been outlined by Mittasch [12].

Most large-scale processes operate today based upon iron catalysts which are similar to those identified originally. In recent decades, an additional process, the Kellogg Advanced Ammonia Process, has been introduced [13]. It is based upon a doubly promoted ruthenium catalyst supported on a hydrogenation-resistant carbon support which is about 20 times more active than the iron-based Haber catalyst. The development of the catalyst, by BP, has recently been described [14]. The catalyst is again structure sensitive, but contrary to the case of the iron catalyst, it comprises dispersed crystallites where alkali promotion facilitates dinitrogen activation via an electronic effect [15]. The active sites for Ru are termed B5 sites and the reaction exhibits pronounced structure sensitivity and particle size dependence [16, 17]. Overall, the uptake of the Ru-based catalytic system on the industrial scale remains low.

Despite the fact that the Haber–Bosch process is highly integrated and energy efficient, there is interest in the further development of ammonia synthesis technology. Currently, the possibility of development of ammonia synthesis in a more sustainable and environmentally friendly manner is attracting attention [18]. The application of hydrogen derived from renewable resources, for example, via electrolytic processes powered by wind energy, offers the prospect of reduction of the significant carbon footprint of the Haber–Bosch process. It is envisaged that such green processes could be operated on more localized scale, facilitating on-demand production of fertilizer close to its point of use, for example, and that the production of ammonia could be applied as an energy store to address periodic oversupply of renewable electricity. In addition, in terms of thermodynamics, the development of highly active catalysts effective at lower reaction temperatures could facilitate production at more localized scale due to the concomitant reduction in the severity of operational parameters. Vojvodic et al. have discussed the possibility of the development of a low-pressure, low-temperature

Haber–Bosch process [19]. Metal-catalyzed ammonia synthesis was described as being limited by the relationship between activation energy for N_2 dissociation and the N-binding energy, the so-called scaling relationship. This limitation arose due to the inability to vary the transition state energy and the N-binding energy to the surface. However, it was argued that there was the possibility of radically different catalysts which could operate under mild conditions which are not subject to the limiting scaling relationship found for metals. To this end, it was suggested that molecular catalysts or catalysts with a limited number of N-binding sites in comparison to the surface metal atoms could prove interesting avenues for exploration since they could facilitate direct N_2 dissociation into a state in which N atoms were singly coordinated.

This chapter details some of the catalysts and approaches which have been detailed for ammonia synthesis within the academic literature within recent years. Whilst most attention is focused upon heterogeneous catalysts, some discussion of the related process of chemical looping is also presented.

7.2 Ruthenium-based catalysts

A number of studies have investigated the role of support for ruthenium-catalyzed ammonia synthesis. Jacobsen has applied boron nitride as a support and has shown that barium-promoted ruthenium is a highly active and stable catalyst [20]. Boron nitride is isoelectronic with carbon and occurs in polymorphs which reflect the carbon allotropes. The BN used in Jacobsen's study possessed a layered structure analogous to that of graphite and it was argued that it did not suffer the disadvantage of the possibility of methanation as would be the case for graphite. The BN-supported catalyst was tested for 3,500 h at 100 bar pressure and 550°C and exhibited no apparent deactivation. Active catalysts can also be prepared using γ-Al_2O_3 as a support [21] and also MgO [21, 22] and $MgAl_2O_4$ [23], supports which comprise promoter elements, have been employed. Recently, Hosono and coworkers have applied electride materials as supports; $Ca_2N{:}e^-$ and $[Ca_{24}Al_{28}O_{64}]^{4+}(e^-)_4$ [24, 25]. The enhanced electron donation power of the resultant catalysts was such that N_2 dissociation was no longer rate determining and inhibition of ammonia synthesis by hydrogen poisoning was reduced due to a reversible hydrogen storage – release process [25]. The surface reactions of N and H adatoms in a Langmuir–Hinshelwood mechanism were reported as rate determining [24]. A schematic of the mechanism over Ru supported on $[Ca_{24}Al_{28}O_{64}]^{4+}(e^-)_4$-related electride is presented in Figure 7.1 [26]. At ca. 2.4 eV, the work function of $[Ca_{24}Al_{28}O_{64}]^{4+}(e^-)_4$ is comparable to that of metallic potassium. The $[Ca_{24}Al_{28}O_{64}]^{4+}(e^-)_4$ electride-supported Ru was be more active than Ru–Cs/MgO, reported to be most active conventionally supported Ru catalyst, exhibiting a turnover frequency a factor of 60 higher at 360°C [27].

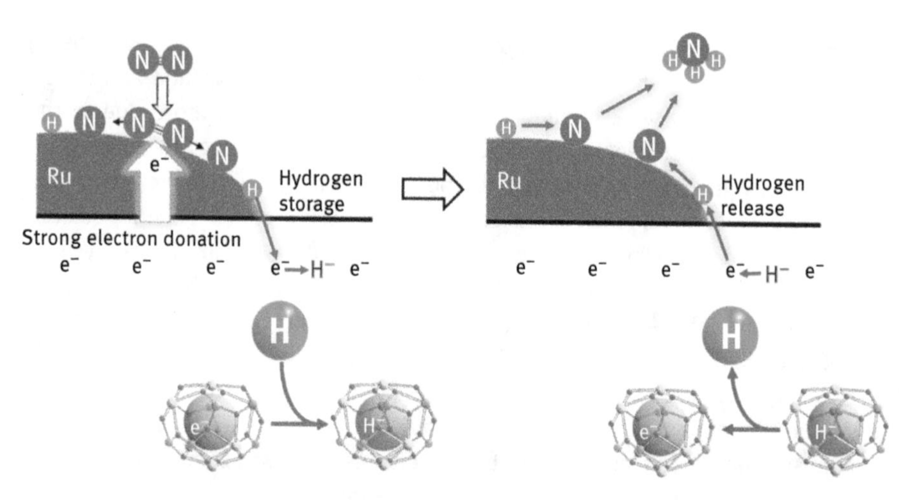

Figure 7.1: Proposed reaction mechanism of ammonia synthesis over $Ru/[Ca_{24}Al_{28}O_{64}]^{4+}(e^-)_4$. Reproduced with permission from Ref. [26]. Copyright 2017 American Chemical Society.

$Ru/Ca_2N{:}e^-$ was reported to be a more active system, with the role of the intermediate hydride formed via the reversible reaction:

$$Ca_2N:e- + xH \rightarrow [Ca_2N]^+ e_{1-x^-} \cdot H_{x^-}$$

being important [28]. It was stated that due to a work function of 2.3 eV, the hydride could facilitate N_2 dissociation. Stable activity at 200°C was evident. Ru/CaH_2 was reported to be as active as $Ru/Ca_2N{:}e^-$. The strength of the Ru–N bond in Ru/Ca_2NH was reportedly crucial for performance with that in Ru/CaNH being weaker and resulting in a catalyst of lower activity [29]. In Ca_2NH, N and H were both anionic whereas in contrast CaNH comprised protons. Related materials have also been investigated as catalysts. Calcium amide has been reported to be an effective support with flat morphology Ru nanoparticles (2.1 ± 1.0 nm) anchored via a strong Ru–N interaction resulting in epitaxial growth of Ru being a very effective catalyst, particularly when promoted with barium which also enhanced stability [30]. At ambient pressure and 340°C, a rate of ca. 12 $mmol.h^{-1}.g^{-1}$ was reported for the Ba-promoted material when applying a stoichiometric ammonia synthesis mixture. Nanosized Ru–Ba core–shell structures self-organized on a mesoporous calcium amide matrix have been reported to be effective catalysts [31].

7.3 Nitrides

It has been reported that, although the nitrogen coverage is low, under industrial conditions, the iron Haber–Bosch catalyst is nitrided in its bulk phase resulting in structural distortion associated with catalytic activity [8]. In early studies, Mittasch drew attention to the nitridation of molybdenum, tungsten, manganese, and uranium during catalysis [12]. In the literature, a number of studies directed at investigation of the ammonia synthesis activity of bulk nitride catalysts have been reported. Amongst systems investigated have been binary nitrides of molybdenum [32–35], uranium [36, 37], vanadium [38, 39], rhenium [40, 41], and cerium [42].

Molybdenum nitrides present an interesting case since in addition to morphology, crystallographic phase can be varied by choice of preparation method [43]. For example, γ-Mo_2N can be prepared by ammonolysis of MoO_3 in a pseudomorphic transformation whereas pretreatment of MoO_3 with 3/1 H_2/N_2 results in the β-$Mo_2N_{0.78}$ phase. In this way, it is possible to probe structure sensitivity. Particle size dependence has been evidenced in the case of γ-Mo_2N with ammonia site time yield ratios at 400°C and ambient pressure of 40:25:1 being reported for particles of size 63 nm, 13 nm, and 3 nm, respectively, being reported [32]. Under similar reaction conditions, the mass normalized activities of γ-Mo_2N and β-$Mo_2N_{0.78}$ were observed to be comparable whereas the δ-MoN phase was found to be inactive [34]. Furthermore, variation of the morphology of the active phases was observed to have little effect upon performance. Transfer of nitrogen from the bulk to the surface was evidenced in the case of Mo_2N in a study in which N_2 activation was reportedly rate determining [44], whereas in a study where the hydrogenation of NH_x species was proposed to be rate determining, the rate of hydrogenation of bulk nitride was reported to be 20–50 times slower than the rate of ammonia synthesis for molybdenum nitride [35]. The hydrogenation of bulk nitride has been reported to be important in the case of uranium nitride [36]. The hydrogenation of lattice nitrogen to form ammonia resulting in transient lattice vacancies which are subsequently replenished from gas-phase nitrogen is termed a Mars–van Krevelen mechanism. Oxygen-based Mars–van Krevelen mechanisms are frequently observed in the case of oxidation reactions catalyzed by metal oxide catalysts [45] but their nitrogen equivalents in the case of nitride catalysts have rarely been considered.

In addition to binary nitrides, more complex ternary nitrides have attracted attention as active ammonia synthesis catalysts. Prominent amongst these have been the Co_3Mo_3N, Fe_3Mo_3N, and Ni_2Mo_3N phases. Co_3Mo_3N has been a particular focus of attention and has been reported to possess higher activity than the iron-based Haber–Bosch catalyst [46–52]. Activity is enhanced by Cs^+ doping, although the addition of too much dopant is found to be detrimental due to issues of phase stability and reduced surface area [49]. The performance of various Co_3Mo_3N-based catalysts benchmarked against a promoted iron catalyst is presented in Table 7.2 [47]. Whilst the reaction conditions applied in the test reported in the table are far from those which would be applied in industrial practice, Co_3Mo_3N-based catalysts have been shown to be of potential interest under conditions at elevated pressure and relating to the presence of a low level of NH_3 in the feed which are more representative [50]. The enhancing role of Cs^+ is argued to be a consequence of a change from NH_2 to NH in terms of the main adsorbed species on the surface and a decrease in

Table 7.2: Ammonia synthesis rates at 400°C and 0.1 MPa pressure, 3/1 H_2/N_2.

Catalyst	Rate ($\mu mol.h^{-1}.g^{-1}$)	Specific activity ($\mu mol.h^{-1}.m^{-2}$)
$Fe-K_2O-Al_2O_3$	330	24
Co_3Mo_3N	652	31
Co_3Mo_3N-K5	869	51
Co_3Mo_3N-K30	364	46
Co_3Mo_3N-Cs2	986	62
Co_3Mo_3N-Cs10	586	59

Table adapted from Ref. [47]. Alkali metal dopants and mol% against Mo shown for doped Co_3Mo_3N catalysts.

strength of NH_x binding [48]. Nitride materials are generally air sensitive, so are commonly stored in passivated form, necessitating *in-situ* reduction to remove the surface oxide layer prior to their application as catalysts. For Co_3Mo_3N, maximum performance was found to result from pretreatment with the reaction mixture for either 12 h at 600°C or for 3 h at 700°C [49].

The performance of the Co_3Mo_3N material has been explained in terms of the relationship between calculated turnover frequency and N_2-binding energy as presented in Figure 7.2 [52]. In this study, it was proposed that an optimum strength of N_2 binding existed. Mo itself bound N_2 too strongly for high activity whereas Co bound N_2 too weakly. However, the average of the two was close to that for Ru, an optimum catalyst. On this basis, the (1 1 1) surface termination plane in which both Co and Mo were expressed was active. The interstitial nitrogen was believed to be non-reactive but of importance in ensuring the correct ordering such that the active surface plane was exposed. This proposal, which is based on theory, is consistent with microkinetic studies performed by the same research group [51]. Inherent in the proposal is the occurrence of structure sensitivity. To the present author's knowledge, structure sensitivity for this catalytic system has not been demonstrated in the academic literature and, indeed, very few studies have been aimed at determination of structure–activity effects in this area. Possibly consistent with the lack of involvement of the lattice N in the catalytic cycle is the observation that Co–Mo alloy nanoparticles, prepared via sodium naphthenide reduction, are active for ammonia synthesis when supported on CeO_2 [53]. The performance of this material was reported to be comparable to that of Co_3Mo_3N and the material comprised ca. 1/20th of the amount of Co and Mo in the bulk nitride. Alternative proposals in which there is a role of reactivity of lattice N in a Mars–van Krevelen-based mechanism for ammonia synthesis have been presented in the literature [54–61]. By substitution of the N_2/H_2 reactant feed by Ar/H_2, low levels of NH_3 production were taken to be indicative of the possibility of active lattice N in the ammonia synthesis mechanism [34, 54, 56]. Furthermore upon high-temperature pretreatment under Ar/H_2, a phase transition occurred in which the transformation of Co_3Mo_3N (which possesses the η-6 carbide structure) into Co_6Mo_6N

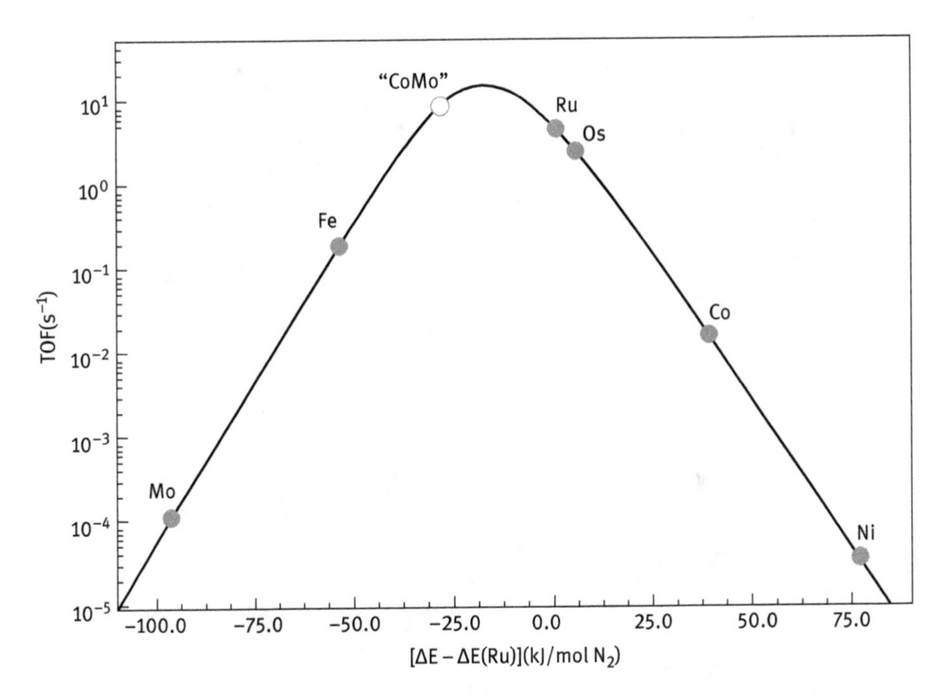

Figure 7.2: Calculated turnover frequencies for ammonia synthesis as a function of the adsorption energy of nitrogen. The synthesis conditions are 400°C, 50 bar, gas composition H_2:N_2 3:1 containing 5% NH_3. Reproduced with permission from Ref. [52]. Copyright 2001 American Chemical Society.

(which possesses the η-12 carbide structure, previously unprecedented for nitrides) occurred [54, 56]. In studies involving *in-situ* powder neutron diffraction, there were no phases of intermediate stoichiometry observed [56]. The process involves the loss of 50% of lattice nitrogen originally present in Co_3Mo_3N and the relocation of the remaining nitrogen from the 16c to the 8a Wyckoff lattice site. Whilst most of the lattice N lost was lost as N_2 rather than ammonia, the process occurs at a significantly higher temperature (700°C) than ammonia synthesis, and rather than being postulated as necessarily part of the reaction mechanism, it is viewed as a demonstration of the reactivity of the lattice N. Further evidence for the reactivity of lattice N in Co_3Mo_3N has been provided in isotopic exchange studies in which the exchange of $^{15}N_2$ with lattice ^{14}N observed [58]. In this study, the role of pretreatment was found to be significant with exchange of 25% of the lattice N being observed after 40 min at 600°C when appropriate pretreatment is applied. Interestingly, the homomolecular exchange of a $^{14}N_2/^{15}N_2$ mixture demonstrated the ability of Co_3Mo_3N to dissociate N_2 to be rather low at ammonia synthesis temperature, despite the material being a good catalyst. It was postulated that the presence of H_2, which was not employed in the reaction, would have a role in the generation of active sites (vacancies) for N_2 activation. This proposal has been supported by computational modeling studies which have been undertaken on surface slabs based upon (1 1 1) termination of the

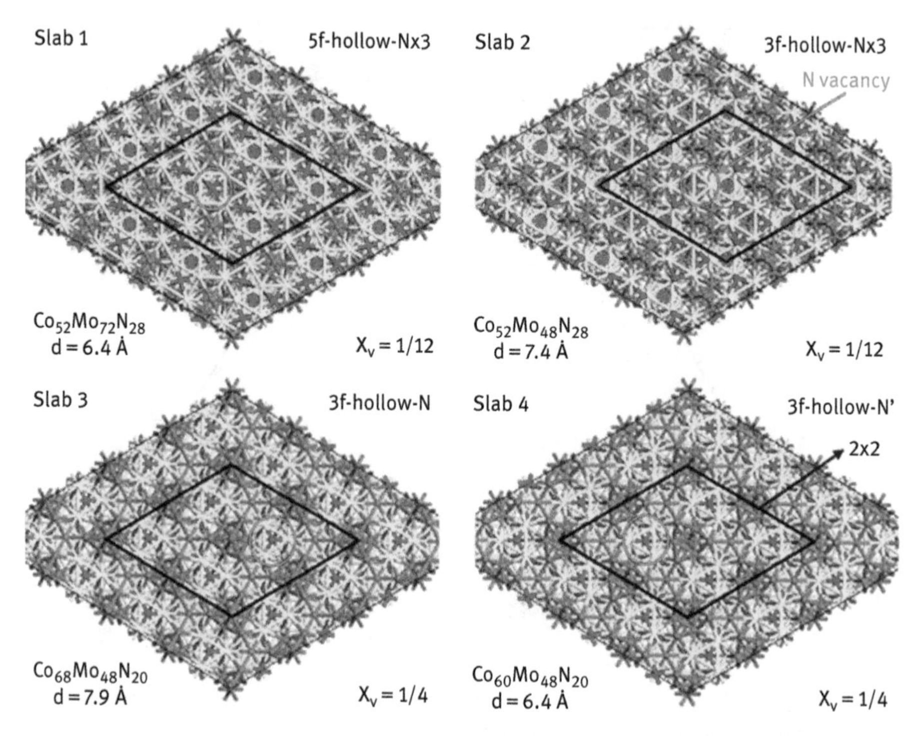

Figure 7.3: 2 × 2 Surface unit cells of the different surface nitrogen-containing (1 1 1) surfaces of Co_3Mo_3N (Co:Mo ≠ 1:1) showing the stoichiometric formula, the slab thickness before relaxation, and the molar fraction of vacancies. The defect-free nitrogen surface concentration of slab 1–2 is three times greater than slab 3–4, and their corresponding nitrogen-vacancy surface molar fractions, Xv, 1 × 1 and Xv, 2 × 2, are shown. The length of the 1 × 1 and 2 × 2 surface vector was 7.832 Å and 15.665 Å, respectively. Mo: yellow; N: blue; Co: green. Reproduced from Ref. [59].

Co_3Mo_3N phase [59]. These slabs are shown in Figure 7.3. The occurrence of both threefold and fivefold N vacancy sites in the surface was proposed and calculations based upon vacancy formation energies demonstrated them to be present in significant concentrations at the temperature range of interest for ammonia synthesis (1.6×10^{16}–3.7×10^{16} cm^{-2} from 380 to 550°C for the threefold sites). Furthermore, N_2 activation as determined by N–N bond elongation for the surface vacancies was reported. In a subsequent study, possible mechanisms involving both the threefold vacancy sites and the cobalt sub-lattice have been reported [60, 61], and a potential low-temperature mechanism, reminiscent of the pathways of biological N_2 activation, has been proposed [61]. In the context of active lattice N, it is interesting to note that Co_3Mo_3C is not active for ammonia synthesis at 400°C [62]. At 500°C, following an induction period wherein partial nitridation of the lattice occurs, ammonia synthesis occurs and continues at steady state during which the carbide is nearly completely nitrided. In addition to being associated with nitridation of the lattice, it is possible that the observed induction period relates either to the reconstruction of the surface or the removal of the surface passivation layer (since the carbide was not pretreated with H_2/N_2 *in situ*, conversion of the carbide to the nitride would occur). It is also not clear whether the presence of lattice N results in the catalytic activity of the material or whether it results from the ammonia produced by the material. Co_6Mo_6C, in contrast, was not active for ammonia synthesis and did not transform to the carbonitride even upon prolonged reaction with H_2/N_2.

In addition to Co_3Mo_3N, Fe_3Mo_3N and Ni_2Mo_3N are also active catalysts for ammonia synthesis [50]. The former is isostructural with Co_3Mo_3N whereas the latter has the filled β-Mn structure. The bulk lattice nitrogen in both phases is less reactive to H_2 than is the case for Co_3Mo_3N [34, 55] and in order to obtain and prepare them caution is necessary to avoid contaminant phases which might be deleterious to performance. Co_3Mo_3N is prepared by controlled ammonolysis of a cobalt molybdate precursor – as dictated by stoichiometry, the analogous procedure in the case of nickel molybdate results in impurity Ni in addition to Ni_2Mo_3N [55] and ferric molybdate results in excess γ-Mo_2N in addition to Fe_3Mo_3N [34]. When the Ni/Mo ratio of the oxide precursor to Ni_2Mo_3N is controlled and prepared via the Pecchini method, Ni_2Mo_3N with enhanced performance can be prepared [63]. In contrast to Co_3Mo_3N, Ni_2Mo_3N can be prepared by direct nitridation of the oxide precursor directly with 3/1 H_2/N_2 which would be preferable to ammonolysis on the large scale due to improved heat management as reported for the preparation of binary molybdenum nitride [64, 65].

Cobalt rhenium is an active catalyst for ammonia synthesis [66, 67]. Whilst Re_3N is active, it deactivates due to partial decomposition into a mixture of Re_3N and Re [67], cobalt rhenium maintains a high performance (492 $\mu mol.h^{-1}.g^{-1}$ at 350°C and ambient pressure with 3/1 H_2/N_2 when the optimal Co:Re 1:4 ratio is applied increasing to 2,372 $\mu mol.h^{-1}.g^{-1}$ at 3.1 MPa). The involvement of a form of rhenium nitride phase as evidenced from XRD was proposed. It was reported that cobalt was necessary to form this nitride phase and rhenium metal and cobalt metal phases were also apparent. Related Ni:Re, Cr:Re, and Cu:Re materials were not active and Fe:Re was not as effective. The inclusion of cobalt was also demonstrated to retard ammonia poisoning upon activity. Whilst the studies detailed have employed an ammonolysis stage, it has been shown that pretreatment with 3/1 H_2/N_2 at 600°C results in active catalysts [68]. Whereas the catalyst prepared by pretreatment with 3/1 H_2/N_2 is immediately active, corresponding pretreatment with 3/1 H_2/Ar results in an induction period which Co and Re edge *in situ* XAS studies indicate to be associated with Co–Re mixing [69]. The difference in pretreatment is also manifested in differences in temperature-programed $^{14}N_2/^{15}N_2$ homomolecular exchange which occurs in Co–Re materials pretreated with 3/1 H_2/N_2 and does not when H_2/Ar is employed. The homomolecular exchange observed in the former case is associated with the desorption of a very low level of H_2 [68]. In addition, TEM evidences different degrees of ordering in the materials as a function of pretreatment [68]. The extent to which the active Co–Re material is nitride is currently not clear. On the basis that the material is generally of very low surface area meaning that its specific activity is very high, enhanced understanding of the origin of its activity could be pivotal in the further development of catalysts of enhanced performance.

7.4 Other catalytic materials

The ternary intermetallic LaCoSi has been reported to be an efficient and stable catalyst for ammonia production [70]. At ambient pressure and 400°C, an ammonia production rate of 1,250 μmol.h^{-1}.g^{-1} has been reported. A combination of experimental and theory-based studies indicated the presence of negatively charged cobalt atoms which promoted activation of N_2, suppressing its desorption, and shifted the dissociation of N_2 as the rate-determining step to the formation of NH_x. Specifically, the mechanism was described as a hot atom mechanism in which the negative cobalt also mediated electron transfer from lanthanum to the adsorbed N_2. The special geometric arrangement of atoms in the material was also proposed to contribute to its activity which was reported to be 60-fold higher than supported cobalt catalysts and also higher than that of Co_3Mo_3N. Related to the LaCoSi system is the use of LaScSi which has been applied as a support for Ru [71]. Similar to the studies described previously [24–28], the support was proposed to have electride-like characteristics with different voids accommodating electrons. Whilst the surface area of the Ru/LaScSi sample was only ca. 2.5 m^2.g^{-1}, it presented an ammonia synthesis rate of 5.3 mmol.h^{-1}.g^{-1} at 400°C and ambient pressure. It was proposed that the origin of the high rate was the ability of the support to absorb and desorb a large amount of hydrogen, also that the support could promote dinitrogen bond cleavage even after absorbing a large amount of hydrogen and that the large number of H^- ions within the LaScSi support and the hydrogen spillover between the Ru and the LaScSi lattice voids means that the dissociated N had a higher probability of associating with H. In terms of highly reactive metals, sodium melt has been reported to be effective for ammonia synthesis [72]. It was shown to be active at 500°C and ambient pressure.

Chen and coworkers have published studies in which they claim to have broken the limiting scaling relationship by applying a two-center reaction pathway [73, 74]. In the first study, a two-center relay pathway was applied which involved the combination of a transition metal and lithium hydride [73]. It was proposed that N_2 was activated by the transition metals and then transferred to LiH where LiNH formed which was able to heterolytically split H_2 forming NH_3 and regenerating the LiH phase. Cr, Mn, Fe, Co, V, and Ni combinations were investigated with the latter two exhibiting poorer activity than the Ru/MgO benchmark material which was applied. Co–LiH and Fe–LiH materials proved to be the best, exhibiting ammonia synthesis rates of around 5,000 μmol.h^{-1}.g^{-1} at 300°C and 10 bar pressure. A kinetic investigation undertaken using a BaH_2–Co/carbon nanotube catalyst indicated that N_2 dissociation was not the rate-determining step in the ammonia synthesis process [74].

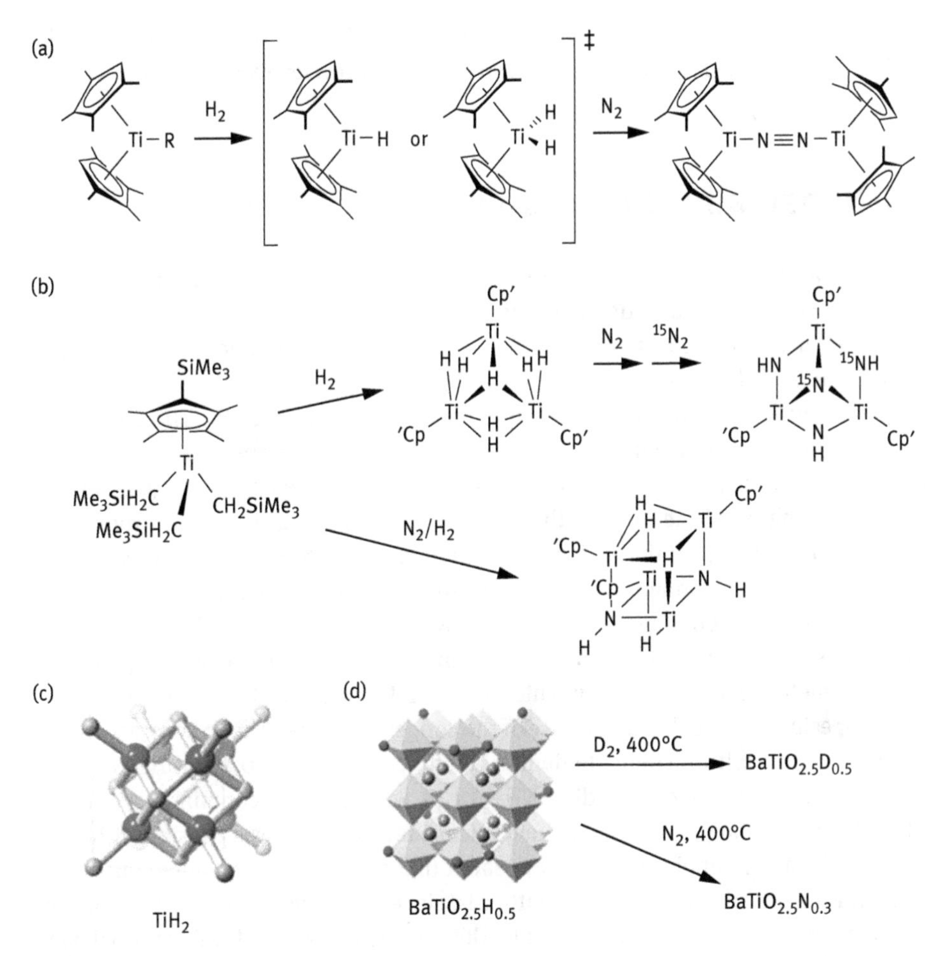

Figure 7.4: Reaction schemes and relevant structures. (a) A titanium-based hydride complex reacting with N_2 gas, (b) a polynuclear titanium hydride complex and its reaction with N_2 gas, (c) the crystal structure of TiH_2 (light blue spheres represent Ti and large blue spheres represent hydrogen), and (d) structure of $BaTiO_{2.5}H_{0.5}$ shown with products from reactions with D_2 and N_2. Reproduced with permission from Ref. [75]. Copyright 2017 American Chemical Society.

Furthermore, activity at a reaction temperature as low as 150°C was apparent [74]. Substitution of BaH_2 with BaO was found to dramatically reduce catalytic performance. Kageyama and coworkers have reported the efficacy of titanium-based hydrides for ammonia synthesis [75]. A schematic showing the various routes for related systems is presented in Figure 7.4.

In the study, TiH_2 and $BaTiO_{2.5}H_{0.5}$ were shown to be active for ammonia synthesis at 400°C and 5 MPa pressure and the formation of an active nitride–hydride surface was proposed [75]. Their activity (up to 2.8 mmol.h^{-1}.g^{-1} for TiH_2 and 1.4 mmol.h^{-1}.g^{-1} for $BaTiO_{2.5}H_{0.5}$) was found to compare to those of Ru/$BaTiO_3$ and Cs-Ru/MgO (at 4.1 mmol.h^{-1}.g^{-1} and 2.7 mmol.h^{-1}.g^{-1}, respectively). The possibility that the hydride catalysts functioned via a rare hydride-based Mars–van Krevelen mechanism was raised.

7.5 Nonsteady state routes

Amariglio and coworkers have reported cyclic operation of ammonia synthesis as a means of overcoming inhibition of the reaction rate by the presence of H_2 for Ru [76] and Os [77] materials. It was proposed that rates of potential commercial significance could be achieved at atmospheric pressure by this approach which involved initial saturation under pure N_2 followed by hydrogenation under pure H_2. The Co_3Mo_3N–Co_6Mo_6N transformation detailed earlier [54–56] affords the opportunity of two-stage ammonia synthesis, especially since it proves possible to regenerate Co_3Mo_3N from Co_6Mo_6N using N_2 alone [57]. In the thermal cycle applied, ammonia production accounts for ca. 8.75% of the lattice N originally in Co_3Mo_3N [54]. The hydrogenation of lattice N to yield ammonia has been further screened in a series of additional nitrides. For the reactive binary nitrides Cu_3N and Ni_3N, up to 30% of their lattice nitrogen is lost as ammonia upon hydrogenation with Ar/H_2 at 250°C [78]. The resultant metals contain porosity which is possibly introduced via the elimination N_2, as especially noticeable in the case of the copper-based system. In contrast, Zn_3N_2 was more stable with incomplete denitridation occurring at 400°C with 23% of the lattice N originally present yielding NH_3. Ta_3N_5 was found to contain less reactive lattice N still, although unlike the other systems, regeneration was possible although conducted via ammonolysis. It also appeared that an amorphous component in the Ta_3N_5 was responsible for the loss of lattice N, since there was no evidence of shifting of reflections in the post-reaction sample. Doping Ta_3N_5 with low levels of cobalt was observed to enhance reactivity at lower temperature [79] and subsequent computational modeling indicated that the cobalt to be located at the nitrogen rich sites and that it enhanced the dissociation of hydrogen as well as lowering the nitrogen vacancy formation energy [80]. Doping has also been shown to enhance the reactivity of manganese nitride materials toward the production of ammonia with 15% of the total available lattice nitrogen being converted to NH_3 for Li–Mn–N at 400°C [81]. The denitridation of the nitrides of iron, cobalt, and rhenium have also been documented [82]. In many of the systems investigated, unlike the case of Co_3Mo_3N, the regeneration of the initial nitride is problematic. This step would be required to develop two-stage reagents for the production of ammonia. In the case of the cobalt molybdenum nitride system, it is likely that incomplete denitridation and strong crystallographic similarity between Co_3Mo_3N and Co_6Mo_6N facilitate the renitridation step.

Pfromm and coworkers have extensively studied systems for the solar thermochemical production of ammonia [83–89]. This approach is designed to develop sustainable fossil-free routes to ammonia. Figure 7.5 presents a schematic which

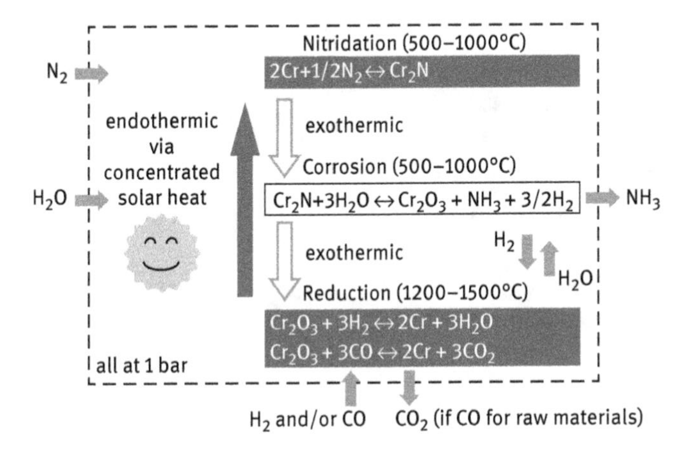

Figure 7.5: Overall approach of N_2 fixation via three-step solar thermochemical NH_3 synthesis at atmospheric pressure. Cr is investigated here for its potential to aid metal oxide reduction and nitridation. Reprinted from Ref. [83] with permission from Elsevier.

details the approach for a chromium nitride-based system [83], which is based upon the following reaction cycle:

$$1/2N_2 + 3H_2O + 3CO \rightarrow NH_3 + 3/2H_2 + 3CO_2$$

$$1/2N_2 + 3/2H_2 \rightarrow NH_3$$

wherein the reducing agent (which can be syngas or hydrogen) is regenerated via endothermic biomass gasification or water dissociation which can both be accomplished by solar processing.

Thermodynamic and economic analyses have shown that the general approach is viable, with indirect CO_2 emissions being in the range 4–50% of those released in the current industrial process employing a coal or natural gas feedstock [84]. The process was also shown to be possibly conceivable for fertilizer production in regions with relatively undeveloped infrastructure. Manganese nitride was reported to be an ideal candidate for the development of ternary nitride redox materials in this context with doping with transition metals being applied to tune lattice nitrogen reactivity [86]. Previously, the nitride iconicity was found to be of importance in the design of solid phase nitride reactants for solar ammonia production with the reduction of H_2O over nitrides yielding NH_3 being governed by the activity of the lattice nitrogen or ion vacancies, respectively [85]. The hydrogenation of alkali and alkaline earth metal nitrides and the reduction of metal nitrides for ammonia production were investigated [87]. Above 550°C at 1 bar, the production of 56.3, 80.7, and 128 μmol NH_3 per mole of metal per minute via reduction of $Mn_6N_{2.58}$ to Mn_4N and hydrogenation of Ca_3N_2 and Sr_2N to Ca_2NH and SrH_2, respectively. The most recently published work

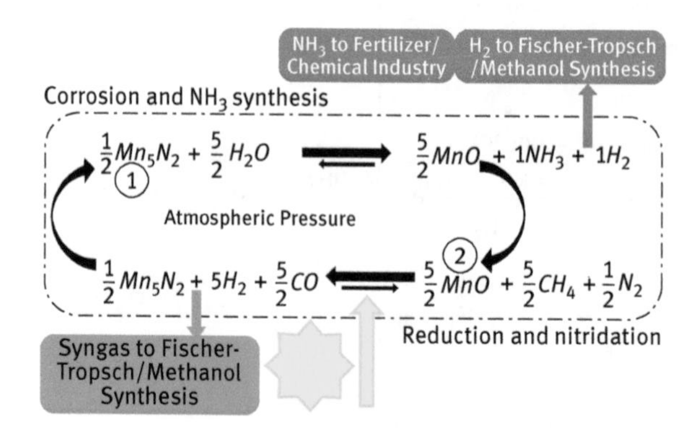

Figure 7.6: Conceptual scheme for NH_3 and syngas (CO and H_2) production via a solar thermochemical reaction cycle at atmospheric pressure. Mn_5N_2 undergoes corrosion with steam in step 1 to produce NH_3 and MnO. The latter is recycled in step 2 via reduction with CH_4 diluted by N_2 to produce syngas and reproduce Mn_5N_2 for reuse in step 1. Reproduced with permission from Ref. [88]. Copyright 2017 American Chemical Society.

from this group in this area has outlined a conceptual scheme for the production of ammonia and syngas from a solar thermochemical cycle employing N_2, H_2O, and CH_4 derived from shale gas [88]. This scheme is presented in Figure 7.6.

Metal nitrides when reacted with protic acids have also been employed as sources of ammonia in organic synthesis. Mg_3N_2 has been employed in this respect [89, 90], although extreme caution related to the possibility of explosion is necessary [91]. AlN has also been applied similarly [92].

Ammonia synthesis from N_2 and H_2O using an electrification cycle in which hydrolysis of Li_3N is a pivotal step has recently been reported [93]. The three steps comprising the process are LiOH electrolysis to yield Li, Li nitridation with N_2, and hydrolysis of Li_3N to regenerate LiOH and liberate NH_3.

7.6 Conclusion

This chapter has presented an overview of some of the recent academic literature pertaining to ammonia synthesis. It is apparent that spurred by the increasing availability of hydrogen derived from renewable resources and also the interest in ammonia as an energy storage vector, there is increasing interest in this topic. The focus of the chapter has been directed toward heterogeneous catalysis and related areas; however, the fields of electrocatalytic and photocatalytic ammonia synthesis are attracting significant degrees of interest, as summarized in a number of recent reviews, for example, Refs. [94–97] and [98, 99], respectively. In addition, the application of plasma-based routes [100] and the influence of electric fields [101] are also topical areas.

As stated in the introduction, the development of catalysts which are active at lower reaction temperatures is of potential interest since advantage could be taken of the more favorable thermodynamic limitations in such regimes, potentially allowing the reduction of process pressure. Robust lower pressure systems which could be run on the small scale and started up and shut down quickly may also drive more localized ammonia synthesis taking advantage of periodic oversupply of sustainable electricity, for example. The localized production of ammonia by such means would also have an additional knock-on effect in terms of reduced requirement for transportation. In relation to the current large-scale centralized Haber–Bosch process, despite the statistics upon energy requirement and greenhouse gas emission, it is important to recognize the degree to which the process is optimized and integrated as well as the catalyst, although very sensitive to poisons, being abundant and long lived. For the large-scale process, it is the hydrogen production which significantly contributes to its energy and greenhouse gas footprint [18]. Whereas studies of materials at ambient pressure and ca. 400°C as frequently reported can have advantages in, for example, investigation of the role of phase transformation upon the development or modification of catalyst performance – and therefore gaining valuable input into catalyst design and optimization, the equilibrium yield of ammonia under such conditions only corresponds to 0.4%. Therefore it would be necessary to further screen materials under more industrially relevant conditions. In looking to the discovery of catalysts active at much lower reaction temperature, lessons can be learned from nature where non-dissociative N_2 activation pathways prevail contrary to the case for the much higher temperature heterogeneous catalytic systems. As is becoming increasingly apparent, computational modeling has a role to play both in terms of rationalizing catalytic performance and also in catalyst design, as exemplified by the recent prediction that Mo doping of Au surfaces could lead to materials active for N_2 dissociation applying near infra-red to visible light nanoplasmonics [102]. In going forward toward the development of sustainable routes to ammonia synthesis, it seems that mutually interdependent multicentered approaches will prove beneficial.

References

[1] Smil V. Detonator of the population explosion. Nature 1999, 400, 415.

[2] Erisman JW, Sutton MA, Galloway J, Klimont Z, Winiwater W. How a century of ammonia synthesis changed the world. Nat Geosci 2008, 1, 636–9.

[3] Smith BE. Nitrogenase reveals its inner secrets. Science 2002, 297, 1654.

[4] Lan R, Irvine JTS, Tao S. Synthesis of ammonia directly from air and water at ambient temperature and pressure. Sci Rep 2013, 3, 1145.

[5] Haber F. The synthesis of ammonia from its elements. Nobel Prize Lecture, 1918. (Accessed 9th February 2018: *https://www.nobelprize.org/nobel_prizes/chemistry/laureates/1918/haber-lecture.pdf*)

[6] Jennings JR. ed. Catalytic ammonia synthesis – fundamentals and practice, Springer, ISBN 978-1-4757-9592-9.
[7] Twigg MV. Ed. Catalyst handbook, Wolfe Publishing Ltd, ISBN 0 7234 0857 2
[8] Schlögl R. Catalytic synthesis of ammonia – A never-ending story? Angew Chem Int Ed 2003, 42, 2004–8.
[9] Somorjai GA, Y. Li Y. Impact of surface chemistry. Proc Natl Acad Sci USA 201, 108, 917–924.
[10] Connor GP, Holland PL. Coordination chemistry insights into the role of alkali metal promoters in dinitrogen reduction. Catal Today 2017, 286, 21–40.
[11] Ertl G, Primary steps in catalytic synthesis of ammonia. J Vac Sci Technol A 1983, 1, 1247–53.
[12] Mittasch A. Early studies of multicomponent catalysts. Adv. Catal 1950, 2, 81–104.
[13] New ammonia process catalyst proven in Canadian plant (Accessed February 2018, http:/www.ogj.com/articles/print/volume-94/issue-47/in-this-issue/refining/new-ammonia-process-catalyst-proven-in-canadian-plant.html)
[14] Brown DE, Edmonds T, Joyner RW, McCarroll JJ, Tennison SR. The genesis and development of the commercial BP doubly promoted catalyst for ammonia synthesis. Catal Lett 2014, 144, 545–52.
[15] Aika K-I. Role of alkali promoter in ammonia synthesis over ruthenium catalysts – effect on reaction mechanism. Catal Today 2017, 286, 14–20.
[16] Jacobsen CJH, Dahl S, Törnqvist E, Jensen L, Topsøe H, Prip DV, Møenshaug PB, Chorkendorff I. Structure sensitivity of supported ruthenium catalysts for ammonia synthesis. J Mol Catal A: Chem 2000, 163, 19–26.
[17] RarÓg-Pilecka W, Mískiewicz E, Szmigiel D, Kowalczyk Z. Structure sensitivity of ammonia synthesis over promoted ruthenium catalysts. J Catal 2005, 231, 11–9.
[18] Pfromm PH. Towards sustainable agriculture: Fossil-free ammonia. J Sust Renew Energ 2017, 9, 034702.
[19] Vojvodic A, Medford AJ, Studt F, Abild-Pederson F, Khan TS, Bligaard T, Nørskov JK. Exploring the limits: A low-pressure, low-temperature Haber-Bosch process. Chem Phys Lett, 2014, 598, 108–12.
[20] Jacobsen CJH. A novel support for ruthenium-based ammonia synthesis catalysts. J Catal 2001, 200, 1–3.
[21] Leterme C, Fernández C, Eloy P, Giagneaux EM, Ruiz P. The inhibitor role of NH_3 on its synthesis process at low temperature over Ru catalytic nanoparticles. Catal Today 2017, 286, 85–100.
[22] Aika K-I., Kumasaka M, Oma T, Kato O, Matsuda H, Watanabe N, Yamazaki K, Ozaki A, Onishi T. Support and promoter effect on ruthenium catalyst. III Kinetics of ammonia synthesis over various Ru catalysts. Appl Catal 1986, 28, 57–68.
[23] Rosowski F, Hornung A, Hinrichsen O, Herein D, Muhler M, Ertl G. Ruthenium catalysts for ammonia synthesis at high pressure: Preparation, characterization and power-law kinetics. Appl Catal A: Gen 1997, 151, 443–60.
[24] Fastrup B. On the interaction of N_2 and H_2 with Ru catalyst surfaces. Catal Lett 1997, 48, 111–9.
[25] Kobayashi Y, Kitano M, Kawamura S, Yokoyama T, Hosono H. Kinetic evidence: The rate-determining step for ammonia synthesis over electride-supported Ru catalysts is no longer the nitrogen dissociation step. Catal Sci Technol 2017, 7, 47–50.
[26] Hara M, Kitano M, Hosono H. Ru-loaded $C_{12}A_7$:e^-electride as a catalyst for ammonia synthesis. ACS Catal 2017, 7, 2313–24.
[27] Kitano M, Kanbara S, Ionue Y, Kuganathan N, Sushko PV, Yokoyama T, Hara M, Hosono H. Electride support boosts nitrogen dissociation over ruthenium catalyst and shifts the bottleneck in ammonia synthesis. Nat Commun 2015, 6, 6731.
[28] Kitano M, Inoue Y, Ishikawa H, Yamagata K, Nakao T, Tada T, Matsuish S, Okoyama T, Hara M, Hosono H. Essential role of hydride ion in ruthenium-based ammonia synthesis catalysts. Chem Sci 2016, 7, 4036–43.

[29] Abe H, Niwa Y, Kitano M, Inoue Y, Sasase M, Nakao T, Tada T, Yokoyama T, Hara M, Hosono H. Anchoring bond between Ru and N atoms of Ru/Ca_2NH catalyst: Crucial for the high ammonia synthesis activity. J Phys Chem C 2017, 121, 20900–4.
[30] Ionue Y, Kitano M, Kishida K, Abe H, Niwa Y, Sasase M, Fujita Y, Ishikawa H, Yokoyama T, Hara M, Hosono H. Efficient and stable ammonia synthesis by self-organized flat Ru nanoparticles on calcium amide. ACS Catal 2016, 6, 7577–84.
[31] Kitano M, Inoue Y, Sasase M, Kishida K, Kobayashi Y, Nishiyama K, Tada T, Kawamura S, Yokoyama T, Hara M, Hosono H. Self-organized ruthenium-barium core-shell nanoparticles on a mesoporous calcium amide matrix for efficient low-temperature ammonia synthesis. Angew Chemie Int Ed 2018, 57, 2648–52.
[32] Volpe L, Boudart M. Ammonia synthesis on molybdenum nitride. J Phys Chem 1986, 90, 4874–7.
[33] Kojima R, Aika K-I. Molydenum nitride and carbide catalysts for ammonia synthesis. Appl Catal A: Gen 2001,219, 141–7.
[34] Mckay D, Hargreaves JSJ, Rico JL, Rivera JL, Sun X-L. The influence of phase and morphology of molybdenum nitrides on ammonia synthesis activity and reduction characteristics. J Solid State Chem 2008, 161, 325–33.
[35] Hillis MR, Kemball C,Roberts MW. Synthesis of ammonia and related processes on reduced molybdenum dioxide, Trans Faraday Soc 1966, 62, 3570–85.
[36] Segal N, Sebba F. Ammonia synthesis catalysed by uranium nitride. 1. Reaction mechanism. J Catal 1967, 8, 105–12.
[37] Segal N, Sebba F. Ammonia synthesis catalysed by uranium nitride. 2. Transient behaviour. J Catal 1967, 8, 113–9.
[38] King DA, Sebba F. Catalytic synthesis of ammonia over vanadium nitride containing oxygen. 1. Reaction mechanism. J Catal 1965, 4, 253–9.
[39] King DA, Sebba F. Catalytic synthesis of ammonia over vanadium nitride containing oxygen. 2. Order-disorder transition revealed by catalytic behaviour. J Catal 1965, 4, 430–9.
[40] Kojima R, Aika K-I. Rhenium containing binary catalysts for ammonia synthesis. Appl Catal A: Gen 2001, 209, 317–25.
[41] Kojima R, Enomoto H, Muhler M, Aika K-I. Cesium promoted rhenium catalysts supported on alumina for ammonia synthesis. Appl Catal A: Gen 2003, 246, 311–22.
[42] Panov GI, Kharitonov AS. Catalytic properties of nitrides in ammonia synthesis. React Kinet Catal Lett 1985, 29, 267–74.
[43] Hargreaves JSJ. Heterogeneous catalysis with metal nitrides. Coord Chem Rev 2013, 257, 2015–31.
[44] Aika K-I, Ozaki A. Mechanism and isotope effect in ammonia synthesis over molybdenum nitride. J Catal 1969, 14, 311–21.
[45] Doornkamp C, Ponec V. The universal character of the Mars and van Krevelen mechanism. J Mol Catal A: Chem 2000, 162, 19–32.
[46] Kojima R, Aika K-I. Cobalt molybdenum bimetallic catalysts for ammonia synthesis. Chem Lett 2000, 514–5.
[47] Kojima R, Aika K-I. Cobalt molybdenum bimetallic nitride catalysts for ammonia synthesis – Part 1. Preparation and characterisation. Appl Catal A: Gen 2001, 215, 149–60.
[48] Kojima R, Aika K-I. Cobalt molybdenum bimetallic nitride catalysts for ammonia synthesis – Part 2 Kinetic study. Appl Catal A: Gen 2001, 218, 121–8.
[49] Kojima R, Aika K-I. Cobalt molybdenum nitride bimetallic catalysts for ammonia synthesis – Part 3 Reactant gas treatment. Appl Catal A: Gen 2001, 219, 157–70.
[50] Jacobsen CJH. Novel class of ammonia synthesis catalysts, Chem Commun 2000, 12, 1057–8.
[51] Boisen A, Dahl S, Jacobsen CJH. Promotion of binary nitride catalysts: Isothermal N_2 adsorption, microkinetic model, and catalytic ammonia synthesis activity. J Catal 2002, 208, 180–6.

[52] Jacobsen CJH, Dahl S, Clausen BS, Bahn S, Logadottir A, Nørskov JK. Catalyst design by interpolation in the periodic table: Bimetallic mmonia synthesis catalysts. J Am Chem Soc 2001, 123, 8404–5.
[53] Tsuji Y, Kitano M, Kishida K, Sasase M, Yokoyama T, Hara M, Hosono H. Ammonia synthesis of Co-Mo alloy nanoparticle catalyst prepared via sodium naphthenide-driven reduction. Chem Commun 2016, 52, 14369–72.
[54] Mckay D, Gregory DH, Hargreaves JSJ, Hunter SM, Sun X-L. Towards nitrogen transfer catalysis: Reactive lattice nitrogen in cobalt molybdenum nitride. Chem Commun 2007, 3051–3.
[55] Hargreaves JSJ, Mckay D. A comparison of the reactivity of lattice nitrogen in Co_3Mo_3N and Ni_2Mo_3N catalysts. J Mol Catal A: Chem 2009, 305, 125–9.
[56] Hunter SM, Mckay D, Smith RI, Hargreaves JSJ, Gregory DH. Topotactic nitrogen transfer: Structural transformation in cobalt molybdenum nitrides. Chem Mater 2010, 22, 2898–907.
[57] Gregory DH, Hargreaves JSJ, Hunter SM. On the regeneration if Co_3Mo_3N from Co_6Mo_6N and N_2. Catal Lett 2011, 141, 22–6.
[58] Hunter SM, Gregory DH, Hargreaves JSJ, Richard M, Duprez D, Bion N. A study of $^{14}N/^{15}N$ isotopic exchange over cobalt molybdenum nitrides. ACS Catal 2013, 3, 1719–25.
[59] Zeinalipour-Yazdi CD, Hargreaves JSJ, Catlow CRA. Nitrogen Activation in a Mars-van Krevelen Mechanism for ammonia synthesis on Co_3Mo_3N. J Phys Chem C 2015, 119, 28368–76.
[60] Zeinalipour-Yazdi CD, Hargreaves JSJ, Catlow CRA. DFT-D3 study of molecular N_2 and H_2 activation on Co_3Mo_3N Surfaces. J Phys Chem C 2016,120, 21390–8.
[61] Zeinalipour-Yazdi CD, Hargreaves JSJ, Catlow CRA. Low-T mechanisms of ammonia synthesis on Co_3Mo_3N. J Phys Chem C 2018, 122. 6078–82.
[62] AlShibane I, Daisley A, Hargreaves JSJ, Hector AL, Laassiri S, Rico JL, Smith RI. The role of composition for cobalt molybdenum carbide for ammonia synthesis. ACS Sust Chem Eng 2017,5, 9214–22.
[63] Bion N, Can F, Cook J, Hargreaves JSJ, Hector AL, Levason W, AMcFarlane AR, Richard M, Sardar K. The role of preparation route upon the ambient pressure ammonia synthesis activity of Ni_2Mo_3N. Appl Catal A: Gen 2015, 504, 44–50.
[64] Wise RS, Markel EJ. Catalytic NH_3 decomposition by topotactic molybdenum oxides and nitrides – effect on temperature programmed gamma-Mo_2N synthesis. J Catal 1994, 145, 335–43.
[65] Wise RS, Markel EJ. Synthesis of high surface area molybdenum nitride in mixtures of nitrogen and hydrogen. J Catal 1994, 145, 344–55.
[66] Kojima R, Aika K-I. Cobalt rhenium binary catalyst for ammonia synthesis. Chem Lett 2000, 29, 912–3.
[67] Kojima R, Aika K-I. Rhenium containing binary catalysts for ammonia synthesis. Appl Catal A: Gen 2001, 209, 317–25.
[68] McAulay K, Hargreaves JSJ, McFarlane AR, Price DJ, Spencer NA, Bion N, Can F, Richard M, Greer HF, Zhou WZ. The influence of pre-treatment gas mixture upon the ammonia synthesis activity of Co-Re catalysts. Catal Commun 2015, 68, 53–7.
[69] Mathisen K, Kirste KG, Hargreaves JSJ, Laassiri S, McAulay K, McFarlane AR, Spencer NA. An *in situ* XAS study of the Co-Re catalyst for ammonia synthesis. Top Catal 2018, 61, 225–39.
[70] Gong Y, Wu J, Kitano M, Wang J, Ye T-N, Li J, Kobayashi Y, Kishida K, Abe H, Niwa Y, Yang H, Tada T, Hosono H. Ternary intermetallic LaCoSi as a catalyst for N_2 activation. Nat Catal 2018, 1, 178–85.
[71] Wu J, Gong Y, Inoshita T, Fredrickson DC, Wang J, Lu Y, Kitano M, Hosono H. Tiered electron anions in multiple voids of LaScSi and their applications to ammonia synthesis. Adv Mater 2017, 29, 1700924.
[72] Kawamura F, Taniguchi T. Synthesis of ammonia using sodium melt. Sci Rep 7 2017, 11578.

[73] Wang P, Chang F, Gao W, Guo J, Wu G, He T, Chen P. Breaking scaling relations to achieve low-temperature ammonia synthesis through LiH-mediated nitrogen transfer and hydrogenation. Nat Chem 2017, 9, 64–70.
[74] Gao, W, Wang P, Guo J, Chang F, He T, Wang Q, Wu G, Chen P. Barium hydride-mediated nitrogen transfer and hydrogenation for ammonia synthesis: A case study of cobalt. ACS Catal 2017, 7, 3654–61.
[75] Kobayashi Y, Tang Y, Kageyama T, Yamashita H, Masuda N, Hosokawa S, Kageyama H. Titanium-based hydrides as heterogeneous catalysts for ammonia synthesis. J Am Chem Soc 2017, 139, 18240–6.
[76] Rambeau G, Amariglio H. Improvement of the catalytic performance of a ruthenium powder in ammonia synthesis by use of a cyclic procedure. Appl Catal 1981, 1, 291–302.
[77] Rambeau G, Jorti A, Amariglio H. Improvement of the catalytic performance of an osmium powder in ammonia synthesis by use of a cyclic procedure. Appl Catal 1983, 3, 273–82.
[78] Alexander A-M, Hargreaves JSJ, Mitchell C. the reduction of various nitrides under hydrogen: Ni_3N, Cu_3N, Zn_3N_2 and Ta_3N_5. Top Catal 2012, 55, 1046–53.
[79] Laassiri S, Zeinalipour-Yazdi CD, Catlow CRA, Hargreaves JSJ. Nitrogen transfer properties in tantalum nitride based materials. Catal Today 2017, 286, 147–54.
[80] Zeinalipour-Yazdi CD, Hargreaves JSJ, Laassiri S, Catlow CRA. DFT-D3 study of H_2 and N_2 chemisorption over cobalt promoted Ta_3N_5-(100),(010) and (001) surfaces. Phys Chem Chem Phys 2017, 19, 11968–74.
[81] Laassiri S, Zeinalipour-Yazdi CD, Catlow CRA, Hargreaves JSJ. The potential of manganese nitride based materials as nitrogen transfer reagents for nitrogen chemical looping. Appl Catal B: Environ 2018, 223, 60–6.
[82] Alexander A-M, Hargreaves JSJ, Mitchell C. The denitridation of nitrides of iron, cobalt and rhenium under hydrogen. Top Catal 2013, 56, 1963–9.
[83] Michalsky, Pfromm PH. Chromium as a reactant for solar thermochemical synthesis of ammonia from steam, nitrogen, and biomass at atmospheric pressure. Solar Energ 2011, 85, 2642–54.
[84] Michalsky R, Parman BJ, Amanour-Boadu V, Pfromm PH. Solar thermochemical production of ammonia from water, air and sunlight: Thermodynamic and economic analyses. Energy 2012, 42, 251–60.
[85] Michalsky R, Pfromm PH. An iconicity rationale to design solid phase metal nitride reactants for solar ammonia production. J Phys Chem C 2012, 116, 23243–51.
[86] Michalsky R, Pfromm PH, Steinfeld A. Rational design of metal nitride redox materials for solar-driven ammonia synthesis. Interf Focus 2015, 5, 20140084.
[87] Michalsky R, Avram AM, Peterson BA, Pfromm PH, Peterson AA. Chemical looping of metal nitride catalysts: Low-pressure ammonia synthesis for energy storage. Chem Sci 2015, 6, 3965–74.
[88] Heidlage MG, Kezar EA, Snow KC, Pfromm PH. Thermochemical synthesis of ammonia and syn gas from natural gas at atmospheric pressure. Ind Eng Chem Res 2017, 56, 14014–24.
[89] Veitch GE, Bridgwood KL, Ley SV. Magnesium nitride as a convenient source of ammonia: Preparation of primary amides. Org Lett 2008, 10, 3623–5.
[90] Bridgwood KL, Veitch GE, Ley SV. Magnesium nitride as a convenient source of ammonia: Preparation of dihydropyridines. Org Lett 2008, 10, 3627–9.
[91] Ley SV, Chemical safety: Mg_3N_2 hazard, Chem & Eng. News 2009, 87, 23, 4.
[92] Ghorbani-Choghamarani A, Zolfigol MA, Hajjami M, Goudarziafshar H,Nikoorazm M, Yousefi S, Tahmmasbi B. Nano aluminium nitride as a solid source of ammonia for the preparation of Hantzsch 1,4-dihydropyridines and bis-(1,4-dihydropyridines) in water via one pot multicomponent reaction. J Braz Chem Soc 2011, 22, 525–31.

[93] McEnaney JM, Singh AR, Schwalbe JA, Kibsgaard J, Lin JC, Cargnello M, Jaramillo TF, Nørskov JK. Ammonia synthesis from N_2 and H_2O using a lithium cycling electrification strategy at atmospheric pressure. Energ Env Sci 2017, 10, 1621–30.
[94] Guo XH, Zhu YP, Ma TY. Lowering reaction temperature: Electrochemical ammonia synthesis by coupling various electrolytes and catalysts. J Energ Chem 2017,26, 1107–16.
[95] Kyriakou V, Gragaounis I, Vasileiou E, Vororros A, Stoukides M. Progress in the electrochemical synthesis of ammonia. Catal Today 2017, 286, 2–13.
[96] Shipman MA, Symes MD. Recent progress towards the electrosynthesis of ammonia from sustainable resources. Catal Today 2017, 286, 57–68.
[97] van der Ham CJM, Koper MTM, Hetterscheid DHG. Challenges in reduction of nitrogen by proton and electron transfer. Chem Soc Rev 2014, 43, 5183–91.
[98] Medford AJ, Hatzell MC. Photon-driven nitrogen fixation: Current progress, thermodynamic considerations and future outlook. ACS Catal 2017, 7, 2624–43.
[99] Chen XZ, Li N, Kong ZZ, Ong WJ, Zhao XJ. Photocatalytic fixation of nitrogen to ammonia: State-of-the-art advancements and future prospects. Mater Horiz 2018, 5, 9–27.
[100] Hong JM, Prawer S, Murphy AB. Plasma catalysis as an alternative route for ammonia production: Status, mechanisms and prospects for progress. ACS Sust Chem Eng 2018, 6, 15–31.
[101] Muakami K, Manabe R, Nakatsubo H, Yabe T, Ogo S, Sekine Y. Elucidation of the role of electric field on low temperature ammonia synthesis using isotopes. Catal Today 2017, 303, 271–5.
[102] Martirez JMP, Carter EA. Prediction of a low-temperature N_2 dissociation catalyst exploiting near-IR-to-visible light nanoplasmonics. Sci Adv 2017, 3, eaao4710.

Index

https://doi.org/10.1515/9783110545210-008

Printed in the USA
CPSIA information can be obtained
at www.ICGtesting.com
LVHW011023041123
762969LV00005BA/37

9 783110 543735